SERVICING THE MIDDLE CLASSES

Domestic service is being reconstituted by the middle classes. Demand for waged domestic labour has escalated rapidly in Britain in recent years. For some, hiring domestic help is nothing new, but for the majority of British households it is an historical fact beyond the range of immediate experience.

Servicing the Middle Classes investigates the resurgence in demand for waged domestic labour in Britain during the 1980s and early 1990s and the consequent growth of a new 'servant' class. Examining the conditions and trends which have conjoined to produce and reproduce this labour force, the book draws extensively on two case studies to explore the geography of demand, day-to-day practices, composition, and social relations of the two major forms of waged domestic labour in contemporary Britain – nannies and cleaners.

The book concludes by considering the effects of the ideologies of motherhood and false kinship relations on the ways in which paid domestic workers are employed, and examines the broader theoretical implications of the study for debates on class and gender and its implications for feminist politics. It will prove of interest to students and researchers working in gender or women's studies, sociology, social policy and social geography.

Nicky Gregson is Lecturer in Geography at Sheffield University.
Michelle Lowe is Lecturer in Geography at Southampton University.

INTERNATIONAL STUDIES OF WOMEN AND PLACE

Edited by Janet Momsen, *University of California at Davis*
and Janice Monk, *University of Arizona*

The Routledge series of *International Studies of Women and Place* describes the diversity and complexity of women's experience around the world, working across different geographies to explore the processes which underlie the construction of gender and the life-worlds of women.

Other titles in this series:

FULL CIRCLES
Geographies of women over the life course
Edited by Cindi Katz and Janice Monk

'VIVA'
Women and popular protest in Latin America
Edited by Sarah A. Radcliffe and Sallie Westwood

DIFFERENT PLACES, DIFFERENT VOICES
Gender and development in Africa, Asia and Latin America
Edited by Janet Momsen and Vivian Kinnaird

SERVICING THE MIDDLE CLASSES

Class, gender and waged domestic labour in contemporary Britain

Nicky Gregson and Michelle Lowe

London and New York

In memory of
'Nanna G' – Rose Gregson (née Taylor), 1899–1980
and
'Grandmother Hibbert' – Gladys Whetton, 1907–1986

First published 1994
by Routledge
11 New Fetter Lane, London EC4P 4EE

Simultaneously published in the USA and Canada
by Routledge
29 West 35th Street, New York, NY 10001

© 1994 Nicky Gregson and Michelle Lowe

Typeset in Baskerville by
J&L Composition Ltd, Filey, North Yorkshire
Printed and bound in Great Britain by
Biddles Ltd, Guildford and King's Lynn

British Library Cataloguing in Publication Data
A catalogue record for this book is available from the British Library

Library of Congress Cataloging in Publication Data
Gregson, Nicky,
Servicing the middle classes: class, gender and waged domestic labour in
contemporary Britain/Nicky Gregson and Michelle Lowe.
p. cm. — (International studies of women and place series)
Includes bibliographical references and index.
1. Domestics—Great Britain. 2. Nannies—Great Britain. 3. Cleaning
personnel—Great Britain. 4. Motherhood—Great Britain. 5. Sex role—Great
Britain. I. Lowe, Michelle, 1963– . II. Title. III. Series: International
studies of women and place.
HD6072.2.G7G74 1994 93–38461
331,4'8164046'0941—dc20 CIP

ISBN 0–415–08530–6
ISBN 0–415–08531–4 (pbk)

CONTENTS

LIST OF FIGURES

LIST OF TABLES

ACKNOWLEDGEMENTS

The book was written without the assistance of any waged domestic labour or study leave! Looking round at the piles of discarded paper which litter bedrooms, living rooms and kitchens; at the mountains of ironing left undone; and at the areas in dire need of 'proper' cleaning, we can see (all too clearly) visible signs of the crisis in daily social reproduction within middle-class households which is the concern of this book. But, whilst we may not have employed others to do our domestic work, there are others without whom this book would not have been written. The project of which this work is part was funded by a two-year research grant from the Economic and Social Research Council (000232817). This financial assistance was vital, and enabled us to employ Sarah James to assist in interviewing, survey work and data collection/analysis. Our particular thanks go too to the secretarial and cartographic staff of the Geography Department of Sheffield University who 'produced' this book – specifically to Kate Schofield (who listened and transcribed for hours and typed forever) and Graham Allsopp; as well as to all those who bore with us, listened and helped us through the summer, autumn and back-end of 1992 when the initial MS was produced – for Nicky, Mave and Man, Sarah, Fred, Caroline, the 'Northern Counties' crew, Doug, Scarlett and three very talented 'WPOs with tails' (Cobber, Ella and Hippy); and for Michelle, Phil, Louise, Ian, Fiona and James. Thanks too to those who helped Nicky condense and revise the original manuscript, particularly to Mike Savage, Nina Laurie and Dina Vaiou who all urged 'say what you really think,' and to Janice Monk and Janet Momsen. However, it is to those who gave so freely and so willingly of their time and thoughts – the subjects of this study – to whom we owe our greatest debt. Those who appear here are all, to some degree, 'in disguise', but they will all – no doubt – recognise themselves and their voices. We owe so much to you all. But it is to Julie, Angie, Nicola and Betty that we owe most. They were the start of it all.

Nicky Gregson, Michelle Lowe

Part I

THE RESURGENCE OF WAGED DOMESTIC LABOUR IN CONTEMPORARY BRITAIN

1

INTRODUCTION
WAGED DOMESTIC LABOUR
IN CONTEMPORARY BRITAIN

Daily nanny (Central Reading) – Sole charge, two children, must be driver/non-smoker.

Smileys Nanny Agency. Permanent and temporary nannies and mothers' helps. Full professional service operated by qualified NNEB. No registration fee to staff.

Nannies where are you? Not going to London or the south-east when there is such a good job on offer here in the north-east. We are looking for a super person used to infants and babies to look after our two children. You will have to be a non-smoker, a car driver, very very reliable and adaptable, smart and intelligent. Your duties are varied and include some evenings and weekends. Travelling abroad and in the UK. Live in or out. Very good negotiable salary and car provided. Newcastle area.

Domestic required. 3 mornings per week. General housekeeping duties. £2 per hour. Jesmond area. References essential.

Domestic cleaner wanted. Approximately 4 hours per week. Chatteris Way, Lower Earley.

Busy Bees offer you a domestic cleaning service. We have no archetypal Mrs Mopp amongst our staff, with nothing more than a steel bucket and a mop of dubious age and origin. Our staff come equipped with modern cleaning equipment and cleaning products plus skills and know-how of when and where and how best to clean.

Each week the national magazine *The Lady* – founded in 1885 and labelled the *Exchange and Mart* of domestic help (Higgins, 1991) – as well as regionally based newspapers, such as the *Newcastle Evening Chronicle*, the *Hexham Courant* and the *Reading Chronicle*, carry advertisements like those above. The advertisements are placed, in the main, by households, although some stem from agencies specialising in the recruitment and supply of various types of domestic staff. Most of these advertisements

3

are for jobs in Britain but others are placed by households living outside Britain and by agencies recruiting for overseas clients. Behind these advertisements are households which are primarily, although not exclusively, middle-class.[1] In most of them, male and female partners are employed, usually full-time, and usually in either professional or managerial occupations. In other cases the female partner may not be in paid employment. In yet other circumstances the advertisement might be placed by a single parent in professional employment. For some of these households hiring domestic help is nothing new. But for the majority of British middle-class households, it is. Domestic service may have been an historical fact but for most of the current generation of employers paid domestic help is a new experience. Instead, for most of the post-war period, domestic labour has been the domain of the full-time unwaged housewife (Oakley, 1974).

Running parallel with the explosion in the small ads have been increasing levels of demand reported by employment agencies specialising in the recruitment and/or supply of domestic staff, and an expansion in the number of firms specialising exclusively in home-cleaning services. The Belgravia Bureau, for example, one of the oldest agencies in London, reported a steady demand from traditional sources for dailies, live-in housekeepers and cooks through the 1980s, but a burgeoning in demand for paid domestic help from younger people (Phillips, 1991). Moreover, in the late 1980s brightly coloured vans (with appropriate 'mop and bucket' or 'Victorian maid' logos, advertising services such as 'The Maids', 'Upstairs Downstairs', 'Poppies' and 'Dial a Char') were to be seen everywhere, particularly in the 'affluent south'. Such vans ferried teams of cleaners, many of them uniformed, to their client middle-class households, there to perform the regular weekly service, spring cleaning or, indeed, pet care, gardening, ironing or even granny minding.

The 1980s in Britain, then, generated some profound changes in the mode of daily social reproduction within a significant number of middle-class households. Not only was domestic labour (or at least certain aspects of this) within these households frequently assuming a waged, as opposed to unwaged, form, but small business capital was expanding into the provision of domestic services. It is this resurgence in waged domestic labour in Britain, specifically its documentation and explanation, as well as the form which this has taken within individual middle-class households, which constitutes the focus for this volume.

Whilst waged domestic labour within middle-class households in contemporary Britain has received little academic attention, it has been the subject of repeated media interest.[2] Stories of, for example, Filipina 'slaves' imprisoned within diplomats' houses in London, and of Middle Eastern and African women entering Britain as the domestic servants

4

of wealthy immigrants, surfaced periodically in the late 1980s and early 1990s.[3] Given their invisible, 'minority interest' subject matter, such stories were not headline news as was, on the other hand, waged domestic labour in relation to childcare. Thus in the summer of 1992 both the quality and the tabloid press devoted considerable space to the kidnapping of baby Farrah Quli by a woman posing as what the press labelled as a 'childminder'.[4] Alongside details of 'the story' were statements from police and representatives of the National Childminding Association and the Professional Association of Nursery Nurses regarding procedures for risk minimisation in the employment of childminders, mothers' helps and nannies. At precisely the same time, the *Beeson* v. *Longcroft* case was being heard in the Old Bailey. Here a twenty-nine-year-old maternity nurse (described variously as either a nurse or a nanny) was cleared of throwing a baby at his mother. However, in the course of the trial the public was treated to predictable suggestions of guilt, accusations of improper motherhood, examples of explosive tensions between nanny and mother and slurs on the professional working mother.[5] In both of these cases reporting made no attempt to stress that such instances represent the exception rather than the norm. Furthermore, no effort was devoted to showing positive examples of childminding or nannying arrangements experienced by many professional working parents. Instead, what we had were high-profile examples of scaremongering and the 'backlash' against those women who seek to combine motherhood and a career (Faludi, 1992; French, 1992). Here then are the journalistic equivalents of the 1992 box office success *The Hand that Rocks the Cradle*, the film which critics suggested would do for professional working mothers what *Fatal Attraction* did for the single career woman.[6]

Waged domestic labour then in contemporary Britain, and particularly that relating to childcare, is an emotive issue. It is a phenomenon which appears to challenge the associations between all women and domestic labour and the assumption that domestic labour is an unwaged activity, carried out for love not money. In addition, it appears to be generative and reflective of some major differences and divisions between women. As we move through this volume it will become clear that such representations, as well as the assumed polarities between the women who work as waged domestics and those who employ them, are grave oversimplifications of an extremely complex phenomenon. Indeed, as we show, waged domestic labour within individual middle-class households in contemporary Britain can only be understood and accounted for in relation to its unwaged form and the ideology and identities which underpin this. Hence, demand for particular categories of waged domestic labour by middle-class households, labour supply, and the social relations of waged domestic labour are shown to be permeated with the unwaged form of domestic labour; with highly traditional ideas about the gendering and form of specific types of domestic labour, and with a strong sense of working for love, as opposed to money.

The research on which this volume is based is part of a larger study which also examines the effects of waged domestic labour on the domestic division of labour within middle-class households with both partners in full-time employment (Gregson and Lowe, 1993, 1994). However, the central concern was with waged domestic labour; specifically its incidence and form within contemporary Britain. The project had three main components. Firstly, we examined the pattern of demand for waged domestic labour within Britain over the ten-year period July 1981 to June 1991, both nationally and within our two case study areas, Newcastle upon Tyne and Reading. Secondly, we surveyed approximately 300 middle-class households in both our study areas in which both partners were in full-time employment in professional and managerial occupations. Such households are the major source of new demand for paid domestic help within contemporary Britain, so this component of our work enables us to discuss the general incidence of waged domestic labour within this particular household form. Finally, through two case studies – conducted in the Reading area of Berkshire in south-east England and the Newcastle and Durham City areas of north-east England – we examined the nature of the nanny and cleaner forms of waged domestic labour. Demand for these two categories of paid domestic help was particularly buoyant through the 1980s.[7] They also represent two different types of domestic work, and, as such, we anticipated that they would constitute two very different forms of waged domestic labour.

The case studies themselves consisted of interviews with 25 nannies and 10 cleaners in both the study areas and of interviews with 29 employers in the north-east and 40 in the south-east.[8] In both cases, 10 matching pairs of nannies and their employers and 5 matching pairs of cleaners and their employers were interviewed. All interviews with nannies and cleaners were conducted either in the homes of their employers (in the employer's absence) or in their own homes. Employers were, for the most part, interviewed at home and were allowed to decide for themselves who should be interviewed. Whilst most of our south-east employing households chose to be interviewed together, the majority of our north-east households decided that this was something for the female partner to deal with![9] The interviews with all categories of respondent were semi-structured, and frequently long and involved. Most interviews with nannies and employers were over an hour in duration, whilst others lasted over two hours. Cleaner interviews tended to be shorter, but some were of an equivalent length and complexity to those with employers and nannies. All interviews were transcribed fully prior to coding and manual analysis.

Implicit within our case study research design was the expectation that there would be significant differences between the two study areas both in the form of waged domestic labour and in its incidence.[10] At the time we anticipated that these differences would reflect the different local economic contexts of both areas in the late 1980s and early 1990s. Typically, our a priori expectations proved to be erroneous. Instead, as we proceeded with our analysis, it became more and more apparent that the differences, such as they were, were slight. Such findings are manifested in the structure of this volume, in which we discuss across space and across our case study areas. As will become clear, it is only where we consider space to matter that we emphasise spatial distinctions and differences. Thus, whilst we show our two local areas to have exerted some influence on waged domestic labour, particularly in relation to labour supply, it is national scale ideological influences and arrangements internal to households which are shown to have been more influential – particularly in producing demand for specific categories of waged domestic labour and in shaping the form which waged domestic labour takes within individual middle-class households.

STRUCTURE OF THE BOOK

Chapter Two is concerned with establishing the pattern of demand for particular categories of waged domestic labour in Britain over the ten-year period 1981 to 1991, and with the pattern within our two case study areas over the same period. This analysis enables us to show the extent of the resurgence in demand for specific categories of waged domestic labour during this period, the geography of this demand and spatial variation. We also show that demand emanated primarily from middle-class households with both partners working in full-time professional and managerial (i.e., 'service class') occupations. In Chapter Three we examine the existing literature on historical and contemporary forms of waged domestic labour in a range of spatial contexts, and the theoretical perspectives which have informed this literature. Although none of these perspectives is central to our account of the resurgence in demand for waged domestic labour in contemporary Britain, some offer useful insights in shaping our explanation of this resurgence. Chapter Three therefore sets up the explanation which is the concern of Chapters Four and Five. In Chapter Four we focus on demand. Here we examine the historical conjuncture which created the potential conditions in which middle-class households with both partners in full-time service class occupations could see the need to substitute particular forms of waged domestic labour for their own unwaged labour. We then look at why those middle-class households whom we interviewed came to employ a nanny, a cleaner, or indeed both forms of waged domestic labour.

Chapter Five takes a similar explanatory form, but is concerned with labour supply. Here we examine the conditions and trends which have conjoined to produce and reproduce a labour force of cleaners and nannies in contemporary Britain. We then illustrate these with reference to the processes through which our particular nannying and cleaning labour forces were produced. Chapter Five ends Part One of this volume, the central concerns of which are to establish and account for the resurgence in waged domestic labour in Britain through the 1980s. In Part Two we examine in depth the day-to-day practices and social relations of the nanny and cleaner forms of waged domestic labour. Chapter Six is concerned with the nanny, Chapter Seven with the cleaner. In each chapter attention is focused on the structuring ideas and principles of these occupations. Of central importance are the ideologies of motherhood and of caring, as well as false kinship relations, all of which have a profound effect on the ways in which our categories of paid domestic worker are employed. Finally, in Chapter Eight we conclude by considering the broader theoretical and political implications of our work.

2

THE GEOGRAPHY OF WAGED DOMESTIC LABOUR IN BRITAIN IN THE 1980s

I originally worked for X Bank, and after I'd had the children I didn't want to return to work for [the bank]. At that time there was a big gap in the market. More and more women were going back to work. I mean, even fifteen years ago, you didn't get the majority of my contemporaries working. And now the majority of my contemporaries do work. There are very few who don't have a job or a career. And there was a big gap in the market. Really there was nothing available in the area. I originally started off with a cleaning side and a nanny side. The nanny side turned in after about two or three years.[1] But the cleaning side went from strength to strength. I've been doing it for about nine years now.

(Domestic Agency, established 1982)

I started with a partner. We started thinking about it just before Christmas 1989. . . . I think a couple of other people had the same idea at the same time because another couple of agencies sprang up. . . . I looked at all sorts of things; generally investigating what was going on in the childcare market, because I felt that we had to be well informed on all kinds of childcare rather than just nannies. . . . We got our licence in May 1990 and started trading in June.

(Nanny Agency, established 1990)[2]

The above comments from two agencies (the first established at the beginning of the 1980s and specialising in household cleaning, and the second founded in 1990 and concentrating on nannies) bear ample testimony to demand for particular categories of waged domestic labour in Britain in the 1980s. However, the agency picture constitutes only one of three elements in the waged domestic labour scene in contemporary Britain. A second is an expanding number of firms specialising in household cleaning services. The third is private hiring: despite the growth of employment agencies and household domestic cleaning services, individual households continue to advertise directly – in regional and national newspapers and magazines, as well as in Job

9

Centres and informally in, for example, Post Office windows – for particular categories of waged domestic labour. In this chapter we concentrate on demand expressed in and through household advertisements. This focus reflects our belief that such demand constitutes the single biggest element in the overall pattern of demand for waged domestic labour in contemporary Britain, and, as such, the most reliable measure of contemporary demand.[3]

The analysis in this chapter has a number of components. Firstly, we examine the national pattern of advertised demand over a ten-year period (July 1981 to June 1991) and the pattern within our two case study areas. We then consider spatial variation in advertised demand; at the county scale within England, within London and within our two study areas. Having established these geographies, we move on to make explicit their association with the spatial distribution of the middle classes. We then show the employment of waged domestic labour in contemporary Britain to be associated predominantly with dual career middle-class households and proceed to establish the incidence of employment of waged domestic labour within such households. Finally, we consider the role of space and place in our conceptualisation of waged domestic labour.

THE DATA

The following analysis is based on three sets of advertisement data. The first comprises data collected from *The Lady* magazine – the major national source of domestic 'sits vac.' – for the period July 1981 to June 1991. The second and third sets, covering the identical period, derive from the two primary local newspapers within our study areas to carry job vacancy listings, namely the *Newcastle Evening Chronicle* and the *Reading Chronicle*.[4] To our knowledge this is the first occasion on which advertisement data have been used as a measure of contemporary demand for waged domestic labour, although such sources have been used in historical studies (see, for example, Cole, 1991 and Gathorne-Hardy, 1972). Because they are related to the circulation of a publication, advertisement data are problematic indicators of demand for waged domestic labour. In our work this problem is compounded by the use of three different sources as well as by our failure to gain some idea of circulation figures for *The Lady*. Furthermore, advertisement data is vulnerable to the problem of variable turnover. Thus, in certain areas – notably London – there is little doubt that at least some households could be advertising annually. Strictly speaking, this would not be a new job but a replacement position. However, we have no way of estimating what proportion of advertisements in our three data sets fell into this category. Given these major data limitations, it is important to stress that the analysis which follows concentrates on establishing broad trends, rather than absolutes.

In the case of the local newspapers, data were collected on a weekly basis, with the day in question being determined by the 'jobs day' of each paper. These data comprised a 100 per cent sample, and are as full a record of household advertisements as is feasible to produce for both our study areas.[5] In contrast, the sheer volume of advertisements carried by *The Lady* meant that we had to sample this source to produce estimates of the total number of household advertisements carried in the study period. The sampling procedure adopted recorded all advertisements occurring in the first week of every other month, again commencing with July 1981. The data on which this part of our analysis is based then approximate to a 12.5 per cent sample of the data source, although not of course of advertisements.[6]

In all cases the data were collected as specified in the advertisement details, before being subjected to preliminary 'cleaning'. All advertisements emanating directly from agencies (or from agencies advertising on behalf of a particular client), were eliminated from further analysis. The remaining advertisements – those from individual households – were then coded according to geographical location. This proved a lengthy process, with varying degrees of success. For our case study areas, the objective was to locate each household advertisement within a local postcode area.[7] In most cases in the north-east this was achieved relatively easily; either because advertisements specified local areas or because the local area telephone code made identification possible. Furthermore, within this region, only a very small proportion of advertisements used box numbers without specifying area. We were therefore able to locate over 90 per cent of all household advertisements occurring in the ten-year study period. In comparison, in the south-east a greater tendency to use box numbers without specifying area, and the existence of the standard trunk dialling (STD) telephone system, meant that we were only able to locate an average of 77 per cent of household advertisements per annum. In the case of data from *The Lady*, and with the exception of London households, we were only aiming for a county location. It therefore proved possible to locate all recorded advertisements. Within London, the tendency for households to specify postcode areas (and the existence of local area telephone codes) made identification at the level of local postcode areas easy. Here over 90 per cent of recorded advertisements were successfully located. Further data 'cleaning' involved two steps. Firstly, the data from our local newspapers were scanned for obvious 'doubles' i.e., advertisements run by the same household on consecutive weeks.[8] Secondly, we checked and (if necessary) reclassified the category of waged domestic labour appearing in each advertisement. Two major areas of reclassification occurred. One

Table 2.1 Recorded and estimated advertised demand for waged domestic labour in Britain: July 1981–June 91

Advertisement	*July 1981*	*1982*	*1983*	*1984*	*1985*	*1986*	*1987*	*1988*	*1989*	*1990*	*June 1991*
Recorded	577	1,201	1,374	1,709	1,742	1,922	1,949	2,176	2,187	1,848	649
Estimated totals	*	9,608	10,992	13,672	13,936	15,376	15,592	17,408	17,496	14,784	*

Total recorded advertisements: 17,334
Estimated UK total, 1981–91: 138,672

Source: The Lady (12.5 per cent sample)
Note: * Estimates have not been produced for 1981 and 1991, given the half-year samples.

involved those advertisements which simply requested 'domestic help'.[9] The other involved the types of differentiation already referred to with respect to the advertisement placed by the Qulis (see p. 294, n. 4).[10]

NATIONAL AND REGIONAL TRENDS: BRITAIN, THE NORTH-EAST AND THE SOUTH-EAST, 1981–91

Table 2.1 records estimated annual totals of advertisements for waged domestic labour placed in *The Lady* by individual households between July 1981 and June 1991. Annual totals climbed steadily through the 1980s, from an estimated low of 9,600 in 1982 to a high of approximately 17,500 in 1989. However, the data for both 1990 and the first half of 1991 suggest a significant contraction.[11]

The number of advertisement categories recorded was a staggering 101 (see Appendix 2.1), including some which harked back to the Victorian/Edwardian era – in name, if not perhaps in nature. Examples were butlers (44 advertisements), governesses (8), maids (28) and foot-men (1). Others, again very much in the minority, were complex amalgamations of activities traditionally gendered as male and female types of domestic work. Instances in this category included advertisements for an 'au pair/driver'; for a 'butler/chauffeur/cook'; and for a 'cook/driver/housekeeper/nanny'. Yet more advertisements combined a number of female activities (for example 'carer/cook/housekeeper' (43 advertisements) whilst others combined a range of male tasks (for example, 'butler/chauffeur' (3 advertisements); 'butler/handyman' (5); 'driver/gardener/handyman' (4) and 'driver/handyman' (5)). However, five categories of waged domestic labour (each accounting for over 5 per cent of recorded demand) accounted for over 74 per cent of recorded advertisements (see Table 2.2). With approaching 30 per cent of recorded advertisements being for nannies, this category was by far the most important of the 101 categories encountered. Moreover, given the nature of the nanny/mother's help and the mother's help jobs,

Table 2.2 Major categories of recorded advertised demand for waged domestic labour in Britain: 1981–91*

Category	Number of advertisements	Percentage
Nanny	4,692	27.1
Mother's help	2,582	14.9
Nanny/mother's help	2,392	13.8
Housekeeper	2,100	12.1
Couples	1,069	6.2

Source: *The Lady* (12.5 per cent sample)
Note: * Major categories are defined as those accounting for >5 per cent of total recorded advertisements.

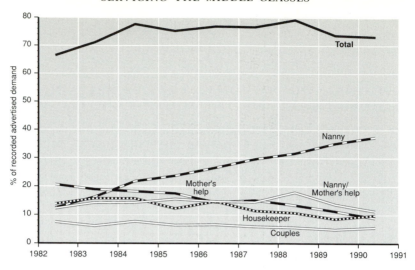

Figure 2.1 Major categories of recorded advertised demand, Britain, 1982–90
Source: *The Lady* (12.5 per cent sample)

it is legitimate to argue that 66 per cent of recorded advertisements related to domestic labour associated with the care of children under 5 years of age. In comparison, the housekeeper category accounted for approximately 12 per cent of recorded advertisements whilst 'couples' – typically a female 'housekeeper' and a male 'gardener/handyman' – accounted for a further 6 per cent.

Figure 2.1 portrays the recorded number of advertisements for nannies, mothers' helps, nanny/mothers' helps, housekeepers and couples for each full year of our study period. Two points are worth highlighting. Firstly, the relative importance of these five categories of waged domestic labour increased from 67 per cent of total recorded advertisements in 1982 to oscillate around 75 per cent of annual totals for the remainder of the study period. Evidently then, their relative importance increased through the 1980s. However, the relative importance of each of these five categories altered through the 1980s. In the early 1980s, for example, it was the mother's help category which was the most important. But, as the 1980s progressed, the relative importance of the nanny category increased, whilst that of the mother's help (and, for that matter, housekeepers and couples) declined. With absolute numbers of advertisements for nannies/mothers' helps, housekeepers and couples staying relatively stable throughout the study period, it is clear that the nanny was not just the most important single category of advertised demand in Britain through the 1980s, but that it became increasingly important as the decade progressed.

Similar patterns, albeit at reduced orders of magnitude, were revealed

Table 2.3 Recorded advertised demand for waged domestic labour in north-east and south-east England, July 1981–June 1991

	July 1981	1982	1983	1984	1985	1986	1987	1988	1989	1990	June 1991
North-east	29 *	35 4.9%**	57 7.9%	68 9.5%	59 8.2%	69 9.6%	57 7.9%	98 13.9%	115 16.0%	98 13.9%	32 *
South-east	46 *	58 5.9%	65 6.6%	88 8.9%	128 13.0%	130 13.2%	116 11.8%	129 13.1%	110 11.2%	87 8.8%	26 *

Sources: Newcastle Evening Chronicle; Reading Chronicle
Note: * Percentage figures have not been produced for 1981 and 1991 given the half-year samples.
　 ** Percentage of total ads in north-east and south-east England.

by the advertisement data collected from our two regionally based newspapers. Table 2.3 details the number of advertisements for waged domestic labour placed by individual north-east and south-east households. Much as with the pattern revealed by advertisements from *The Lady*, the number of advertisements placed in both areas climbed steadily through the early years of the 1980s, and then expanded dramatically. In the case of the south-east, this increase occurred in the mid 1980s. However, in the north-east, the expansion seems to have occurred later, with the years 1988–90 witnessing numbers roughly twice those of previous years. Both areas suggest a rapid fall off in the number of advertisements for 1991 – again a pattern which mirrors that suggested by data from *The Lady*.

Appendix 2.2 details the pattern of advertised demand for all categories of waged domestic labour in both study areas through the 1980s. In the north-east, six categories of waged domestic labour were revealed: the nanny, the nanny/mother's help, the mother's help, the cleaner, the housekeeper and the au pair. 34 per cent of all north-east household advertisements were for nannies; 25 per cent for cleaners; 18 per cent for mothers' helps; 13 per cent for nannies/mother's helps; 10 per cent for housekeepers and 1 per cent for au pairs. Here then, much as nationally, it is waged domestic labour in relation to the care of young, pre-school-age children which comprised the dominant category of advertised demand. Indeed, 47 per cent of all north-east household advertisements were concerned with seeking waged domestic labour to take *sole* charge of pre-school-age children – a figure which is very much in accordance with our national findings.

Percentage figures across the ten-year period show a marked change in the nature of advertised demand from north-east households through the 1980s. Again this resonates with the national picture. Thus, in the early years of the 1980s the dominant category was that of cleaners but in the second half of the 1980s the percentage share of cleaner advertisements, although not actual totals, declined considerably. The reason for this was the increase in the number of nanny advertisements placed by north-east households, particularly in the period 1988–90. The expansion in advertised demand for waged domestic labour by north-east households through the 1980s, then, was closely bound up with an increase in demand for nannies in particular.

In the south-east, the overall picture is, on first appearances, rather more complex. Rather than advertised demand being concentrated in six categories, in the south-east it fell into nineteen. Of these, the most important was the mother's help (27 per cent), followed by the nanny (14 per cent), the cleaner (14 per cent), the housekeeper (13 per cent), the gardener (9 per cent) and the nanny/mother's help (7 per cent). The remaining categories, however, all accounted for rather marginal percentages. Despite initial appearances to the contrary, then, advertised demand for waged domestic labour in the south-east through the 1980s – as nationally and as in the north-east – was closely associated with a limited range of categories. Unlike the case with both the national and the north-east data sets, though, only 21 per cent of this advertised demand can be directly related to waged domestic labour for the *sole* charge of pre-school-age children; although if we include the mother's help category, this figure rises to a more comparable 48 per cent.

Percentage figures for each category in the south-east across the ten-year study period reveal similar trends to those nationally and in the north-east. During the early 1980s the dominant categories were those of housekeepers and mothers' helps – both features which replicate the national pattern. However, during the second half of the 1980s the percentage share of housekeepers declined considerably, whilst both the nanny *and* the cleaner categories gained in relative importance. As nationally and in the north-east, the declining importance of house-keepers did not reflect an absolute decline in levels of recorded advertised demand. Instead, it reflected an increase in the number of advertise-ments for nannies and cleaners. It is, however, noticeable that in the south-east the percentage share of advertisements accounted for by the nanny category, even in the later years of the 1980s, was not as great as in either the national or the north-east data sets. Moreover, by 1991 only six of the nineteen recorded categories of advertised demand were still 'active'.

In summary, the aggregate pattern of recorded advertised household demand revealed by all three sets of advertisement data for the period July 1981 to June 1991 is remarkably consistent. Indeed, it is possible to identify four main characteristics. These are

1 An expansion in advertised demand for waged domestic labour in Britain through the 1980s, with demand increasing considerably, both nationally and within our two study areas, through the second half of the 1980s.
2 A concentration of demand in a small number of categories, particularly in those categories pertaining to the care of young, pre-school-age children and to general housework.
3 A change in the dominant categories of advertised demand through the 1980s. Both nationally, and within our two study areas (although to a lesser extent in Berkshire than in the north-east), it is waged domestic labour in relation to the care of young pre-school-age children which currently dominates. However, in the early 1980s demand was primarily for housekeepers and mothers' helps.
4 A possible decline in advertised demand in 1991 – although, given the limited nature of our data here, this can only be speculative.

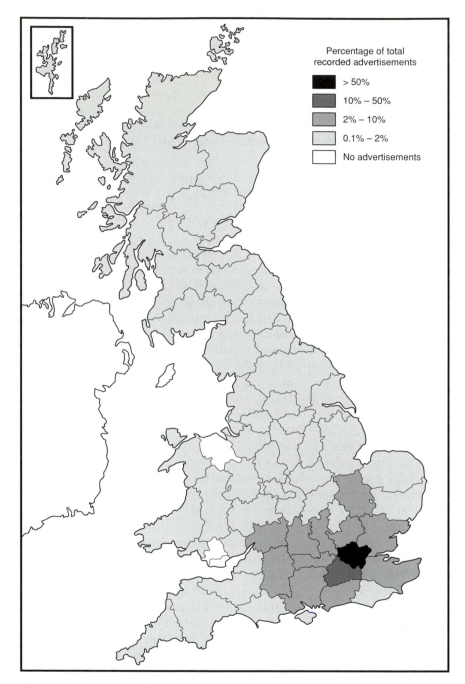

Figure 2.2 Spatial distribution of recorded advertised demand, Britain, 1982–91
Source: *The Lady* (12.5 per cent sample)

Such findings confirm the observational and anecdotal comments referred to in Chapter One. With total recorded advertised demand increasing nationally and within our case study regions, it is clear that it is appropriate to talk about the resurgence in demand for waged domestic labour in contemporary Britain. However, this resurgence would appear to have been connected primarily with the care of young, for the most part, pre-school-age children.

Having outlined general aggregate patterns we now move on to look at spatial variation in advertised household demand.

SPATIAL VARIATION IN ADVERTISED HOUSEHOLD DEMAND: 1981–91

Figure 2.2 portrays advertised household demand for all categories of waged domestic labour over the period July 1981 to June 1991 for all counties in Britain. The pattern suggested is one of spatial concentration. Demand appears to have been concentrated overwhelmingly in the 'affluent south', particularly in London and certain of the home counties. In contrast, the north appears as a sea of non-existent or insignificant demand. Indeed, Durham and Tyne and Wear rank 37th and 38th respectively in the overall county listing of advertisements (see Appendix 2.3). Given the total number of north-east advertisements (Appendix 2.2), it would be inaccurate to infer from Figure 2.2 that demand for waged domestic labour outside the 'affluent south' was insignificant. Instead, we are evidently dealing with a source biased towards London, the home counties and the south. However, the overall picture revealed by Figure 2.2 is, almost certainly, correct. Advertised household demand for waged domestic labour in Britain through the 1980s was concentrated predominantly in London, the south and the south-east regions.

Excluding London, which we consider separately later, this spatial pattern has two important components. The first is the importance of Surrey as a source of demand for waged domestic labour. Indeed, Surrey accounted for 25 per cent of the British (excluding London) data set (with the second ranked, Berkshire, at 8 per cent). A second identifiable component is that a further set of counties constituted a clear secondary concentration of recorded advertised household demand. These were: Berkshire (8 per cent); Hertfordshire (6.5 per cent); Hampshire (6.1 per cent); Kent (7.6 per cent); West Sussex (5.3 per cent) and Buckinghamshire (5.3 per cent) (see also Appendix 2.4).

Figure 2.3 reveals the pattern of advertised household demand for Surrey, Berkshire, Buckinghamshire, Oxfordshire and West Sussex (i.e., the five counties with significantly large location quotients) through the 1980s. The overall trend here is much as indicated both by figures for Britain (Table 2.1) and for our case study areas (Table 2.3). However,

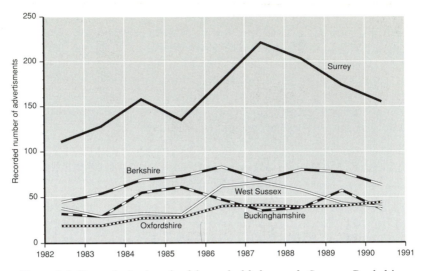

Figure 2.3 Recorded advertised household demand: Surrey, Berkshire, Buckinghamshire, Oxfordshire, West Sussex, 1982–90
Source: *The Lady* (12.5 per cent sample)

Table 2.4 Major categories of recorded advertised demand, England (excluding London), 1981–91*

Category	Number of advertisements	Percentage
Nanny	1493	18.0
Housekeeper	1462	17.7
Mother's help	926	11.2
Nanny/mother's help	835	10.1
Couples	617	7.5
Cook/housekeeper	534	6.5
Carer/housekeeper	429	5.2

Source: *The Lady* (12.5 per cent sample)
Note: * Major categories are defined as those accounting for ⩾5 per cent of total recorded advertisements

there are also important differences between the five counties. The case of Surrey is the most dramatic. Here recorded advertised household demand escalated rapidly in the period 1985–7 and, since then, has declined. This pattern is one which replicates almost exactly that revealed by advertisements from the *Reading Chronicle*. In comparison, the increase in demand for the remaining counties (with the exception of Oxfordshire) was much more gradual, peaking in all cases in the mid 1980s. With Oxfordshire, demand increased through the mid to late 1980s and remained relatively buoyant even in 1990. This has greater similarities with the pattern revealed by advertisements from our north-east study area.

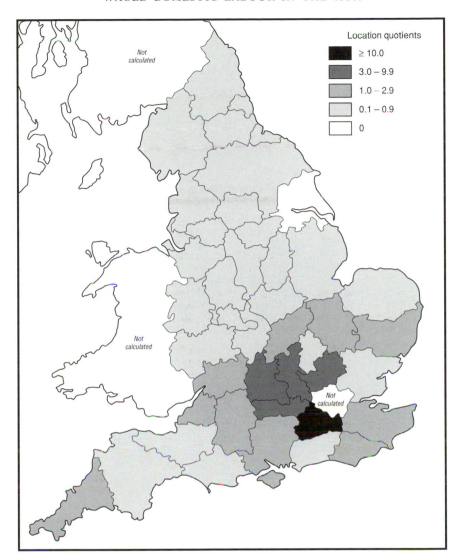

Figure 2.4 Location quotients for nanny advertisements: all English counties and
metropolitan areas (excluding London), 1981–91
Source: *The Lady* (12.5 per cent sample)

Appendix 2.4 details recorded advertised demand for each category
of waged domestic labour within the English counties. Since many of
these categories are numerically insignificant we confine attention here
to the major categories only. As might be anticipated, given the dominance of Surrey within the overall county picture, there is only one major
category in which this county does not rank first. However, it is in

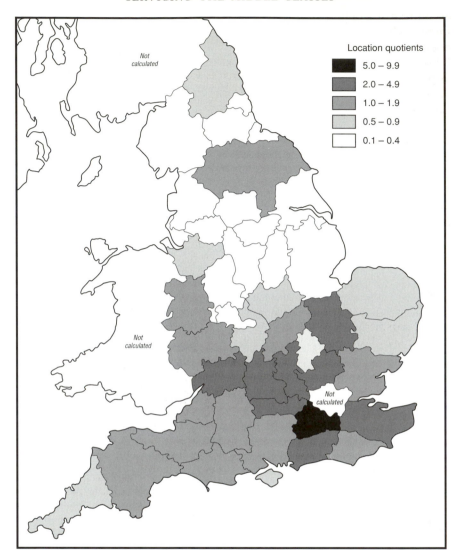

Figure 2.5 Location quotients for housekeeper advertisements: all English
counties and metropolitan areas (excluding London), 1981–91
Source: *The Lady* (12.5 per cent sample)

relation to demand pertaining to childcare that Surrey's dominance is
most marked – notably in connection with nannies, nannies/mothers'
helps and mothers' helps. Thus Surrey accounted for 25 per cent of all
nanny advertisements, for 28 per cent of nannies/mother's help advertisements and for 26 per cent of all mother's help advertisements. In the
case of all three of these categories, Berkshire, Kent and Hampshire each

accounted for between 7 per cent and 9 per cent of recorded advertisements, whilst Hertfordshire and Buckinghamshire each accounted for over 5 per cent. Thus, six counties accounted between them for nearly 62 per cent of recorded advertised household demand for nannies; for approximately 66 per cent of that for mothers' helps and for 68 per cent of that for nannies/mothers' helps. In contrast, the patterns revealed by categories relating to general housework were rather different. Surrey itself only accounted for 14 per cent of all recorded advertisements for housekeepers; for 15 per cent of those for couples and for 13 per cent of those for cook/housekeepers. Moreover, the list of dominant counties for general housework categories includes those counties isolated previously, as well as others. Figures 2.4 and 2.5 present the geographies of advertised demand for nannies and housekeepers in Britain through the 1980s.

In summary then, and in spite of clear data limitations beyond the south/south-east, county level advertisement data suggest:

1 The concentration of recorded advertised household demand for waged domestic labour in the 'affluent south' and south-east in the 1980s – particularly within Surrey, Berkshire, Buckinghamshire, Oxfordshire and West Sussex.
2 An increase in recorded advertised demand within these areas through the middle years of the 1980s. This appears to have been associated closely with advertised demand in relation to childcare, specifically that relating to pre-school-age children.
3 The spatial concentration of advertisements seeking waged domestic labour in relation to childcare in the 'new money' counties surrounding London, and the broader spatial base of advertised demand for general housework categories (in which both the 'new money' and 'shire' counties figure).

Having established the broad patterns of recorded advertised demand within Britain, we move next to examine the patterns within London itself.

As suggested by Figure 2.2, London households constituted the primary source of advertised demand for waged domestic labour in Britain in the 1980s. Indeed, they accounted for 52 per cent of all recorded advertisements. However, the relative importance of London

Table 2.5 London advertisements: percentage share per annum, 1982–90

1982	1983	1984	1985	1986	1987	1988	1989	1990
567	683	849	918	972	1,002	1,201	1,237	1,016
47.2	49.7	49.7	52.7	50.6	51.4	55.2	56.5	55.0

Source: *The Lady* (12.5 per cent sample)

Table 2.6 Major categories of recorded advertised demand, London, 1981–91*

Category	Number of advertisements	Percentage
Nanny	3,199	35.2
Mother's help	1,656	18.2
Nanny/mother's help	1,557	17.1
Housekeeper	638	7.0

Source: The Lady (12.5 per cent sample)
Note: * Major categories are defined as those accounting for ⩾5 per cent of total recorded advertisements

Table 2.7 Major categories of recorded advertised demand by postcode area, London 1981–91*

NW**	SW	W	N	SE	WC	E	EC	Total
1996	2,635	1,020	912	306	107	62	12	7,050
28.3%	37.8%	14.5%	12.9%	4.3%	1.5%	0.9%	0.2%	

Source: The Lady (12.5 per cent sample)
Notes: * Major categories as in Table 2.2.
  ** *North-west*: Brondesbury, Camden Town, Childs Hill, Cricklewood, Golders Green, Hampstead, Hampstead Garden Suburb, Harlesden, Hendon, Kensal Green, Kentish Town, Kilburn, Kingsbury, Neasden, Primrose Hill, Queensbury, Regent's Park, South Hampstead, St John's Wood, Stonebridge, The Hyde, West Hampstead, West Hendon, Willesden, Willesden Green.
South-west: Balham, Barnes, Battersea, Brixton, Castenail, Chelsea, Clapham, Earlsfield, East Sheen, Fulham, Mortlake, Pimlico, Putney, Roehampton, South Kensington, South Lambeth, Stockwell, Streatham, Wandsworth, Westminster.
West: Acton, Bayswater, Chiswick, Ealing, East Acton, Hammersmith, Kensington, Maida Vale, Notting Hill, Paddington, Shepherds Bush.
North: Crouch End, East Finchley, Fortis Green, Highbury, Highgate, Holloway, Hornsey, Hoxton, Islington, Muswell Hill, Stoke Newington, Stroud Green, Tottenham, Upper Holloway.
South-east: Bermondsey, Blackheath, Brockley, Bushey Green, Camberwell, Catford, Deptford, Dulwich, East Dulwich, Eltham, Forest Hill, Greenwich, Grove Park, Herne Hill, Kennington, Lambeth, Lee, Lewisham, Lower Sydenham, Nunhead, Rotherhithe, Upper Sydenham, Walworth, Woolwich.
East: Aldersbrook, Beckton, Bethnal Green, Bow, Bromley, Canning Town, Dalston, East Ham, Forest Gate, Hackney, Hackney Wick, Haggerston, Higham Hill, Homerton, Isle of Dogs, Leabridge, Leyton, Leytonstone, Limehouse, Lower Clapton, Manor Park, Mile End, Millwall, Old Ford, Plaistow, Poplar, Shoreditch, Silvertown, Snaresbrook, Stepney, Stratford, Upper Clapton, Upton Park, Walthamstow, Wanstead, Wapping, West Ham, Whitechapel.
West Central: Bloomsbury, Charing Cross, Covent Garden.
East Central: Barbican, Bishopsgate, Clerkenwell.

households as a source of recorded advertised demand increased through the 1980s (Table 2.5). Thus, whilst in 1982 London households accounted for 47 per cent of all recorded advertisements, by 1989 this had risen to nearly 57 per cent. Beyond this, it is important to note that the range of important categories of advertised demand within London was confined to four (i.e., the nanny; the mother's help; the nanny/mother's help and the housekeeper), with the nanny accounting for 35 per cent of all London advertisements (Table 2.6). Given the importance of the nanny, mother's help, nanny/mother's help and housekeeper categories within London, we confine our examination of the spatial pattern of recorded advertised demand within London to these four categories (see Appendix 2.5 for a full record of the spatial location of all categories of recorded advertised demand).

There is a distinctive (and predictable) geography to advertised demand for waged domestic labour within London (Table 2.7). Thus, in general terms, it is the south-west and north-west postcode areas which emerge as the dominant areas of advertised demand. Indeed, between them these two areas accounted for over 66 per cent of recorded advertisements. Against this, the north and west postcode areas – with 13 per cent and 15 per cent of recorded advertisements respectively – appeared as secondary areas of demand, whilst the remaining postcode areas proved to be of minimal importance. However, there is evidence to suggest that the geography of advertised demand for waged domestic labour within London changed through the 1980s (Figure 2.6). Thus, as the 1980s progressed, the importance of the north-west postcode areas declined whilst that of the south-west increased. Clear differences also emerged between areas in terms of dominant categories of advertised demand. So, although the south-west postcode area represented the biggest source of advertised demand (Table 2.8), accounting for approaching 39 per cent of all nanny advertisements in London in the 1980s, for nearly 40 per cent of

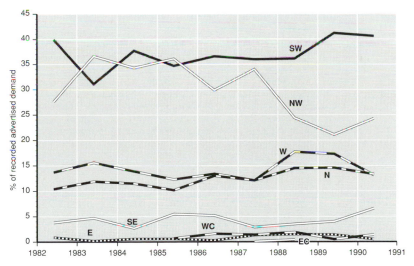

Figure 2.6 Percentage share of recorded advertised demand by postcode area, London, 1982–90
Source: *The Lady* (12.5 per cent sample)

Table 2.8 Demand for nannies, mothers' helps and housekeepers by postcode area, London, 1981–91

Category	NW	SW	W	N	SE	WC	E	EC	Total
Nanny	736	1,239	514	440	172	55	39	4	3,199
	23.0%	38.7%	16.1%	13.8%	5.4%	1.7%	1.2%	0.1%	
Mother's help	555	608	188	213	54	24	11	3	1,656
	33.5%	36.7%	11.4%	12.9%	3.3%	1.4%	0.7%	0.2%	
Nanny/	450	620	201	187	69	17	10	3	1,557
mother's help	28.9%	39.8%	12.9%	12.0%	4.4%	1.1%	0.6%	0.2%	
Housekeeper	255	168	117	72	11	11	2	2	638
	39.9%	26.3%	18.3%	11.3%	1.7%	1.7%	0.3%	0.3%	
									7,050

Source: *The Lady* (12.5 per cent sample)

Table 2.9 Recorded percentage annual advertised demand for nannies, mothers' helps and housekeepers by postcode area, London, 1982–90*

Postcode Area	Category	1982	1983	1984	1985	1986	1987	1988	1989	1990
NW	**1	24.4	32.5	27.3	33.7	25.1	27.2	20.1	17.3	19.2
	2	28.6	38.0	33.3	39.7	34.2	40.6	36.5	26.3	31.3
	3	23.9	34.4	34.4	33.5	33.6	38.3	20.8	22.9	28.3
	4	36.4	45.3	56.9	41.1	31.6	43.5	33.3	36.7	45.5
SW	1	31.9	31.1	39.4	35.2	36.6	36.0	37.6	43.4	42.9
	2	44.9	28.8	37.9	34.8	36.4	38.8	30.9	43.4	34.3
	3	45.8	40.8	38.6	39.1	37.4	37.2	40.1	38.3	42.8
	4	27.3	20.0	25.3	21.9	35.0	26.1	28.0	25.0	27.3
W	1	17.0	17.2	16.4	11.8	16.2	15.7	19.3	16.9	13.3
	2	14.0	15.3	14.1	11.8	9.2	4.6	11.0	13.2	11.8
	3	7.3	12.0	11.7	8.9	9.6	11.0	17.8	19.8	13.0
	4	18.2	18.7	10.1	23.3	22.8	14.5	22.7	21.7	14.5
N	1	12.8	11.9	13.8	11.1	14.2	12.0	14.9	15.0	15.0
	2	6.2	14.7	11.6	7.3	14.7	13.1	15.5	15.1	14.7
	3	14.6	7.2	10.4	11.7	11.2	10.5	12.9	13.4	9.4
	4	12.7	13.3	6.3	10.9	7.0	14.5	13.3	13.3	5.5
SE	1	6.4	6.6	2.2	7.0	5.9	4.5	3.8	4.8	7.8
	2	2.2	3.1	2.5	4.9	3.8	1.7	2.8	0.6	5.9
	3	5.2	5.6	4.3	5.6	6.4	1.7	4.9	3.9	4.3
	4	1.9	2.7	1.3	1.4	1.8	–	–	3.3	3.7

Source: *The Lady* (12.5 per cent sample)
Notes: * The E, EC and WC areas have been omitted due to their overall insignificance as sources of advertised demand. Percentages may not therefore add to 100.
    ** 1 Nanny
       2 Mother's help
       3 Nanny/mother's help
       4 Housekeeper

advertisements for nannies/mothers' helps, and for 37 per cent of advertisements for mothers' helps, the north-west accounted for a greater share of housekeeper advertisements (40 per cent, as against 26 per cent). We can also note that whereas the percentage share of annual advertised demand for housekeepers accounted for by the north-west and south-west areas changed little through the 1980s, this was not so with demand in the nanny category. Here the relative importance of the south-west increased (1982: 32 per cent; 1990: 43 per cent) (Table 2.9).

When we look at what was happening within London's postcode areas through the 1980s, the same pattern of spatial concentration is to be found. Chelsea – with over 10 per cent of located London advertisements – proved the most important general source of advertised demand in the 1980s (Figure 2.7). Other important sources of demand were Clapham (6.2 per cent), Hampstead (9.4 per cent), Islington (8.2 per cent) and Kensington (7.7 per cent). These areas, together with Chelsea, accounted for 44 per cent of all London advertisements.

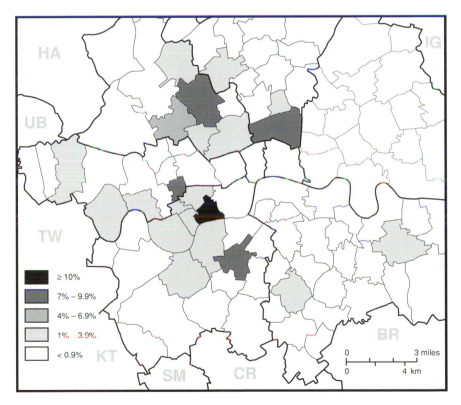

Figure 2.7 Spatial distribution of advertised demand for waged domestic labour, London, 1982–90
Source: *The Lady* (12.5 per cent sample)

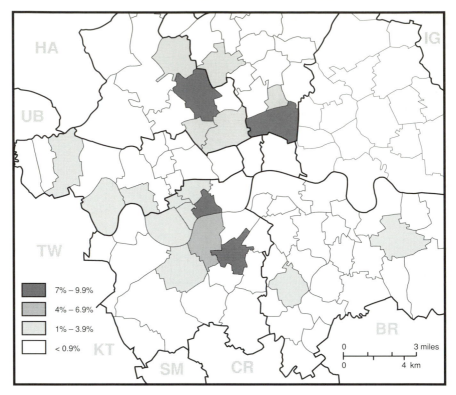

Figure 2.8 Spatial distribution of advertised demand for nannies, London,
1982–90
Source: *The Lady* (12.5 per cent sample)

For the most part, the patterns of advertised demand for both nannies
and housekeepers through the 1980s echo this general picture (Figures
2.8 and 2.9) but there are some important differences. So, whilst Chelsea
emerges as an important area of advertised demand in the nanny
category (7.6 per cent of all such advertisements in London in the
1980s), it is not the most significant source of advertised demand.
Instead, Islington and Clapham – each with 8 per cent of total advertisements – constitute the primary sources of demand in this category. As
with the general picture, Hampstead (7.7 per cent) and Kensington (6.8
per cent) are of considerable importance too, but it is the south-west
postcode area which emerges as an area of relatively high levels of
advertised demand in the nanny category. In this area Chelsea (7.6 per
cent), Clapham (8.0 per cent), Battersea (4.2 per cent) and Wandsworth
(3.5 per cent) all recorded over 3 per cent of the total recorded for
London (and together make up 23 per cent of the London total).

Looking at the housekeeper category, two areas dominate the overall

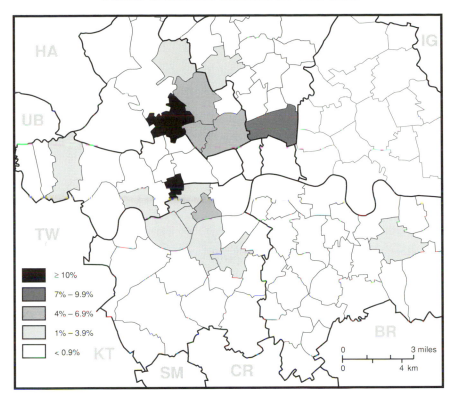

Figure 2.9 Spatial distribution of advertised demand for housekeepers, London, 1982–90
Source: The Lady (12.5 per cent sample)

picture (South Hampstead and Kensington). Moreover, although Islington emerges as another important source of advertised demand, a different set of areas features prominently in the housekeeper category. Thus, along with Chelsea and Hampstead, Regent's Park and St John's Wood prove important sources of advertised demand for housekeepers. The south-west areas – which feature so strongly in the overall pattern of advertised demand for nannies within London – are of less importance. In contrast, four areas within the north-west postcode area (South Hampstead, Hampstead, Regent's Park and St John's Wood) account for nearly 28 per cent of the London total in the housekeeper category.

The conclusion to be drawn from Figures 2.7–2.9 is that advertised demand, both for waged domestic labour generally in London and for particular categories of waged domestic labour through the 1980s, was spatially concentrated – not just in the south-west, north-west, west and north postcode areas, but within these areas too (Figures 2.10, 2.11, 2.12, 2.13). In general, it is Hampstead, Chelsea, Clapham, Kensington

29

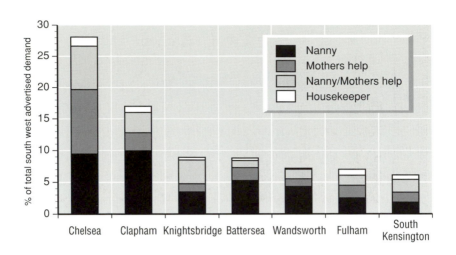

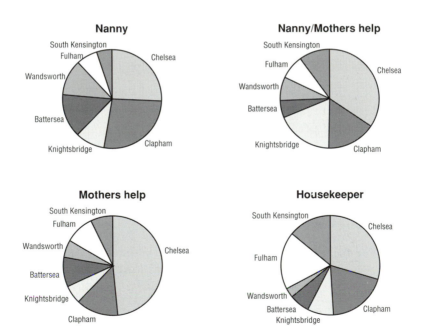

Figure 2.10 Advertised demand for waged domestic labour, south-west London, 1982–90

Source: *The Lady* (12.5 per cent sample)

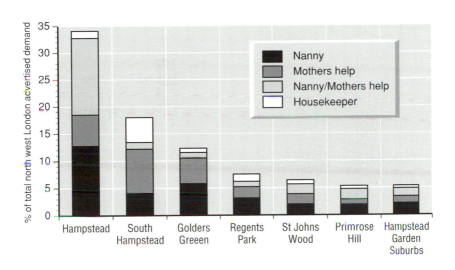

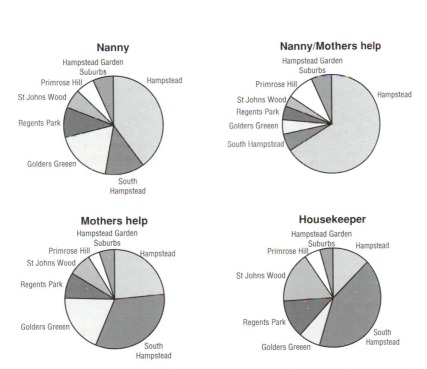

Figure 2.11 Advertised demand for waged domestic labour, north-west London,
1982–90
Source: *The Lady* (12.5 per cent sample)

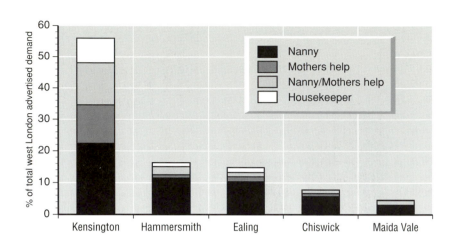

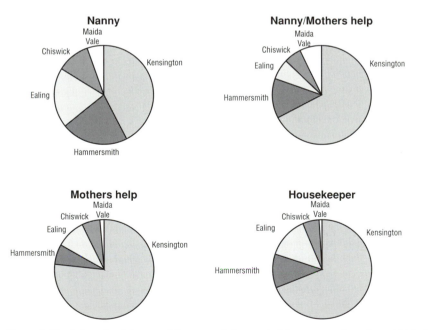

Figure 2.12 Advertised demand for waged domestic labour, west London, 1982–90
Source: The Lady (12.5 per cent sample)

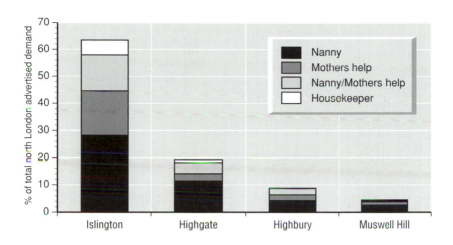

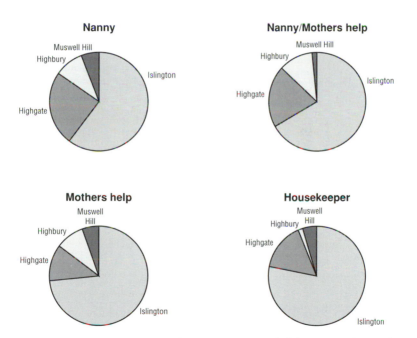

Figure 2.13 Advertised demand for waged domestic labour, north London,
1982–90
Source: *The Lady* (12.5 per cent sample)

and Islington which emerge as areas of particular concentration. However, there is no doubt too that the geography of advertised demand for nannies in London in the 1980s differs in some respects, both from the general picture and from the specific geography of advertised demand for housekeepers. Thus, it is the south-west areas of Chelsea, Clapham, Battersea and Wandsworth which constituted the spatially extensive 'nanny country' of London in the 1980s, whilst the housekeeper equivalent comprised the north-west areas of South Hampstead, Hampstead, Regent's Park and St John's Wood.

Unlike our national data, the spatial specificity of our London data set enables us to say rather more about the types of areas behind these advertisements, although not – of course – about those households advertising for such forms of labour. Not surprisingly, our maps of advertised demand for waged domestic labour within London coincide closely with maps of two other phenomena – owner-occupation and professional/managerial employment.[12] As comparisons of the 1981 Census with that of 1971 showed, large percentage point increases in owner-occupation occurred over this period in some inner London boroughs, and particularly in Hammersmith, Fulham, Kensington and Chelsea (Congdon, 1987: 12), whilst above average increases in the percentage of professional/managerial workers occurred in Hammersmith, Fulham, Wandsworth, Islington and Haringey (Congdon, 1987: 31–2). It is precisely these areas, along with the established high status central areas of Kensington and Chelsea, which feature in our data set, either as areas of consistently high demand for waged domestic labour or as areas of increasing importance. Moreover, these same areas, particularly Hammersmith, Fulham, Kensington and Chelsea, are shown to have been those in which large percentage point increases occurred between 1971 and 1981 in the numbers of 25–44-year-olds (Congdon, 1987: 12), i.e., those age groups in the 'family formation' life-cycle stage. Given the increasing importance of single-person households in this age group, it cannot be assumed that such areas are dominated by conventional household forms, but it would seem reasonable to suggest that these are areas characterised by significant numbers of households in the 'family formation' life-cycle stage, many of whom would be potential nanny employers. The major sources of advertised demand for waged domestic labour within London then are those areas associated with above average levels of 25–44-year-olds, with above average levels of owner-occupation and with above average levels of professional/managerial employment. Undoubtedly, they are also areas associated with above average levels of female participation in full-time waged work (see, for example, Duncan, 1991: 425). At the same time we can note that the geography of demand in the housekeeper category is almost certainly indicative of the older, established middle-class nature of north-west London.

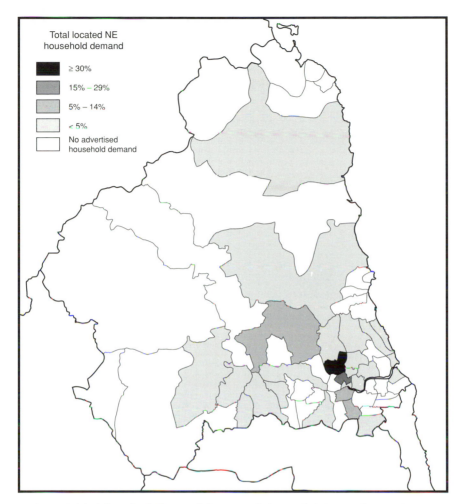

Total located NE
household demand

≥ 30%

15% – 29%

5% – 14%

< 5%

No advertised
household demand

Figure 2.14 Advertised demand for waged domestic labour, the north-east,
1981–91
Source: *Newcastle Evening Chronicle*

Much as with the London advertisement data, both the north-east and
south-east newspaper surveys revealed distinctive patterns of spatial
concentration. In the north-east it was overwhelmingly the Gosforth and
Jesmond areas of Newcastle (as well as, although to a much lesser
degree, Ponteland/Darras Hall, Low Fell and Whitley Bay) which were
behind the advertised demand for waged domestic labour within the
Tyneside conurbation throughout the 1980s (Figures 2.14 and 2.15).
Both the South Gosforth and Jesmond wards are characterised by above
average owner-occupation rates and well above average male and female

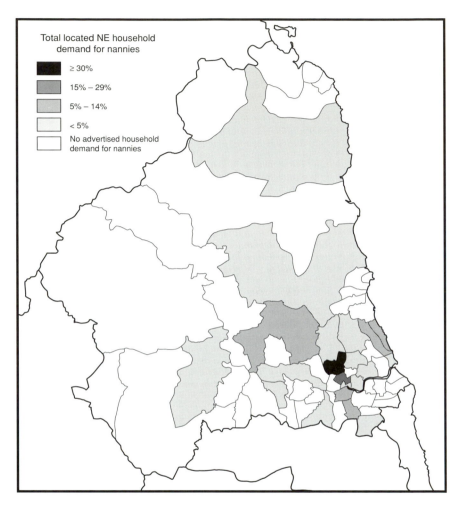

Figure 2.15 Advertised demand for nannies, the north-east, 1981–91
Source: *Newcastle Evening Chronicle*

employment levels.[13] They also show low levels of overcrowding and low levels of most measures of social deprivation. Such a picture is confirmed by occupation data from the 10 per cent census sample. Of employees living in South Gosforth, 40 per cent were in professional and/or managerial occupations in 1981, whilst 33 per cent of those Jesmond residents in employment were in similar occupations. This compares with a figure of 14 per cent for the city as a whole. The Gosforth and Jesmond areas then − areas associated with the urban professional middle classes − are those which lie behind advertised demand for waged domestic labour in the north-east.

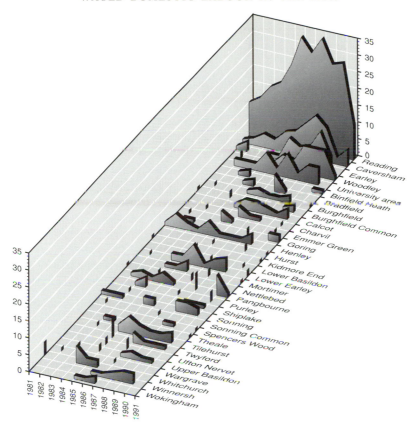

Figure 2.16 Advertised demand for waged domestic labour, Reading area, 1981–91

Within our south-east case study area an examination of the spatial pattern of located advertisements for waged domestic labour through the ten-year study period revealed a more complex pattern than in the north-east (Figure 2.16). Here we were unable to locate a consistently dominant source of recorded advertised demand. But a shifting geography of demand through the ten-year period could be identified. In the early years of the 1980s, advertisements were associated primarily with the belt to the north of Reading (principally in south Oxfordshire, and encompassing a number of 'Thameside' settlements such as Henley on Thames and Whitchurch, as well as a large number of villages in the Chilterns). By the mid 1980s there were a number of other concentrations, notably the relatively established middle-class areas of Woodley and Tilehurst, together with the 'new middle-class' suburban estate of Calcot. However, by the later 1980s these two areas had been joined by two further new middle-class housing estates, Lower Earley and

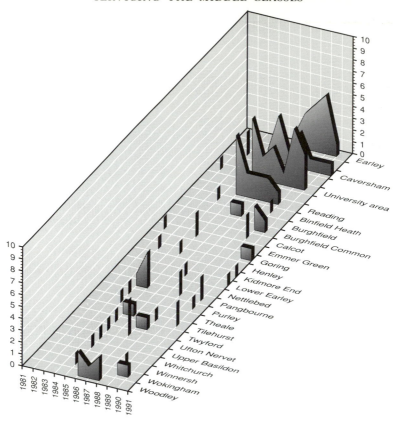

Figure 2.17(a) Advertised demand for nannies, Reading areas, 1981–91

Winnersh. Much as in the north-east, the general pattern of advertised demand was echoed by the pattern of demand in the nanny category (Figure 2.17).

Notwithstanding these spatial shifts, two places in particular stand out as consistently visible centres of advertised demand in the south-east study area through the 1980s, namely parts of Caversham (Thames ward) and the relatively prosperous 'university area' (notably Redlands ward) (see Figure 2.16). Thames ward has a high percentage of owner-occupied housing and one of the highest percentages of professional/ managerial occupations in the Reading Borough. Census data thus suggest that the socio-economic characteristics of this area of Caversham are extremely similar to those of Gosforth. The 'university area' of Reading comprises three main wards (Park, Redlands and Church) of which Redlands is the most likely source of advertised demand for waged domestic labour. With above average levels of owner-occupation and approximately 30 per cent of those employed being in professional/

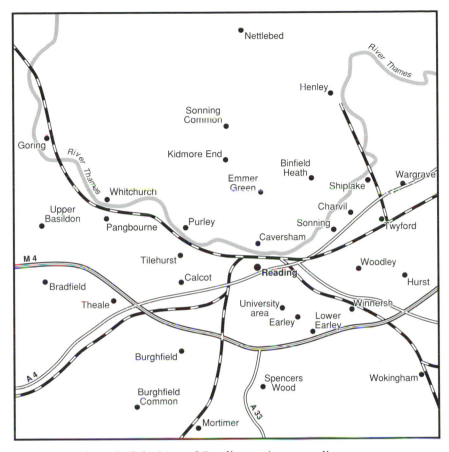

Figure 2.17(b) Map of Reading and surrounding areas

managerial occupations, this area seems to correspond closely to the equivalent 'university area' in our north-east case study (i.e., Jesmond).

The above geographies of advertised demand for waged domestic labour – be they at the national scale or within different conurbations – are ones which show a high degree of correspondence with high levels of owner-occupation, professional/managerial employment and female full-time working in paid employment (see, for example, Thrift (1987: 233); Duncan (1991: 425) and Savage *et al.* (1992: 164)). Assuming a correspondence between advertised demand for waged domestic labour and the employment of waged domestic workers, 'waged domestic labour country' therefore is associated with the distribution of the middle classes in contemporary Britain. However, the geographies of advertised demand for different categories of waged domestic labour are not necessarily identical. As the London analysis shows, it is –

inevitably – the geography of the younger middle classes (ie., the 25–44 age group) with which the geography of advertised demand for nannies coincides. In London through the 1980s, as research on gentrification particularly has shown, it was the south-west areas (Clapham, Battersea, Wandsworth, Fulham, etc.) which became associated with this segment of the middle classes, leaving the north-west area as that associated with the established (and older) middle classes and demand in the house-keeper category (Congdon, 1987; Hamnett, 1976; Hamnett and Williams, 1980; Munt, 1987). In the Reading area much the same process seems to have occurred, as is evidenced by the shifting geography of advertised demand for nannies from Caversham and the south Oxfordshire area to include areas associated with the new middle classes. In contrast to this, in the north-east it is stability which is the overall picture.

Having established the association between the geography of advertised demand for waged domestic labour and the geography of the middle classes, we move on to consider the association between waged domestic labour and the middle classes of contemporary Britain in more depth.

WAGED DOMESTIC LABOUR AND THE MIDDLE-CLASSES IN CONTEMPORARY BRITAIN

Although the similarities between the spatial distribution of the middle classes and advertised demand for waged domestic labour do not imply a causal link between the two phenomena, we are in no doubt that the two are strongly related and that the geography of the one is but a manifestation of the geography of the other. Confirmation for this causal connection came from pilot survey work, as well as from interviews with employment agencies and with those working as waged domestics. In 1990 we conducted a pilot survey of 'users' of waged domestic labour in both our case study areas.[14] The respondents were accessed via employment agencies specialising in matching domestic staff and client households. One of the agencies was located in Newcastle, the other in Reading. In both areas waged domestic labour was employed predominantly in dual-career households, within which both partners were employed full-time in professional/managerial occupations. These findings were later confirmed by interviews with employment agencies and waged domestics.

Having established the association between advertised demand for waged domestic labour and the middle classes in professional and managerial employment, a second stage of our research was to produce estimates of the incidence of use of waged domestic labour amongst middle-class dual-career partnerships. In order to achieve this we surveyed a total of 542 dual-career partnerships (with both partners in full-time employment in professional/managerial occupations) in Reading and Newcastle (see Appendix 2.6). Of the 268 households in the north-east with both partners in full-time employment in professional/ managerial occupations, 97 (36 per cent) employed waged domestic labour in one form or another. In the south-east, 91 (33 per cent) of 274 dual-career households employed waged domestic labour. For the most part households were employing one person. However, in the case of the north-east, 17 households (18 per cent) employed more than one person, whilst in the south-east 20 households (22 per cent) employed more than one individual. The incidence of employment of waged domestic labour within middle-class dual-career households then would seem to be consistent across space. Indeed, our findings suggest that over a third of such households in contemporary Britain employ waged domestic labour in one form or another.[15]

Table 2.10 shows the forms of waged domestic labour employed by our two samples. Again the pattern is one of consistency across space. Thus, in both the north-east and south-east, it was the cleaner category of waged domestic labour which proved to be by far the most commonly employed. Indeed, in both areas, over 70 per cent of the dual-career households surveyed which employed waged domestic labour were employing a cleaner.[16] In comparison, approximately 20 per cent of these households employed a nanny and around 15 per cent a gardener, with the remaining categories being insignificant.

Whilst the percentage figures in Table 2.10 provide a reliable guide to the relative incidence of employment of cleaners and gardeners within dual-career households, they considerably under-estimate the incidence of nanny employment. This is because the vast majority of dual-career households only employ a nanny in relation to care of pre-school-age children. When we take into account the presence/absence of

Table 2.10 Incidence of employment of categories of waged domestic labour in case study areas

Category	North-east		South-east	
	Nos employed (n = 103)	Percentage of households employing these categories	Nos employed (n = 104)	Percentage of households employing these categories
Cleaner	67	72.8	67	76.1
Nanny	15	16.3	18	20.5
Gardener	14	15.2	12	13.6
Mother's help	4	4.3	1	1.1
Au pair	1	1.1	4	4.5
Other	2	2.2	2	2.2

Source: Workplace survey data
Note: Percentages are calculated from 92 households in the NE and 88 households in the SE due to incomplete returns.

pre-school-age children within those households surveyed and employing waged domestic labour, the incidence of nanny employment increases considerably. Thus, of the 97 north-east households employing waged domestic labour, 30 had one or more pre-school-age children and, of these, 14 (47 per cent) employed a nanny. In the south-east, of the 91 employing households, 38 had one or more children in this age range, 14 (37 per cent) of whom employed a nanny. Such findings lead us to suggest that approaching three-quarters of all dual-career households employing waged domestic labour employ a cleaner and that between a third and a half of such households with pre-school-age children employ a nanny. They also suggest that the cleaner and the nanny are the major forms of waged domestic labour used by the middle classes in contemporary Britain. We examine why this is so in Chapter Four.

Having outlined the geography of contemporary waged domestic labour in Britain and its close association with the middle classes, particularly with dual-career households, in the final section of this chapter we consider the role of space and place in our conceptualisation of waged domestic labour.

SPACE, PLACE AND WAGED DOMESTIC LABOUR

Although advertisements for waged domestic labour in Britain through the 1980s reveal a clear geography, and although we take this geography (and the spatial distribution of the middle classes which it reflects) to be of importance in understanding the match between demand and labour supply, space and place do not figure centrally in our conceptualisation of contemporary waged domestic labour.

Earlier we made explicit that the differences between our north-east and south-east case study areas were slight. We also stated that this apparent lack of importance of place was difficult to make sense of, at least initially. Although the bias in *The Lady* as a data source means that it is important to be sceptical about the suggested pattern and level of advertised demand for waged domestic labour outside London and the south/south-east regions, we are in no doubt about the concentration of the middle classes in these regions and the implications of this for waged domestic labour. This spatial concentration (and the greater numbers of waged domestics and employing households which this implies in the south-east region) was something which, at the start of our work, we anticipated would be influential. Being more widespread and more visible within the south-east region than in the north-east, we assumed that there would be greater levels of awareness of waged domestic labour as a local social practice within middle-class households in our south-east case study area. In turn, we envisaged that this would make a difference to the phenomenon of waged domestic labour in the south-east as compared to that in the north-east. Initially we anticipated finding differences in, for example, social relations; in how and/or why households came to be employing waged domestics; even perhaps in the types of people whom they employed. Moreover, we also anticipated that the differences between the two local labour markets in the 1980s would be of some significance.

Such thinking is reflective of the space/place (geography) matters theme, and particularly of its expression in the notion of the causal potential of local areas (see, for example, Massey, 1984 and the work which this inspired: Bagguley *et al.*, 1990; Cooke, 1985, 1989; McDowell and Massey, 1984; cf. Duncan and Savage, 1989, 1991). However, the more we looked at waged domestic labour, the more convinced we became that it was the similarities in all this which mattered. Indeed, it really didn't seem to matter whether we were talking to a couple of systems analysts in Reading or a pair of senior grade local government workers in Newcastle, nannies in Durham or nannies in Lower Earley, cleaners at work in Gosforth or cleaners in Caversham – the points which they made, the relations which they described and the ideas which they articulated were all interchangeable across space and between place. Moreover, as we show in Chapters Four, Six and Seven, neither the hows and/or whys of employing waged domestics, nor the social relations of waged domestic work had anything remotely to do with local areas and/or local social practices. Furthermore, as our survey work shows, there appears to have been little difference between our two study areas in the percentage use of either waged domestic labour generally or of particular categories of waged domestic labour amongst dual-career households. In such circumstances we were forced to conclude that here was a phenomenon which, at least in the context of contemporary Britain, demonstrates that geography (in the form of place specific social practices) was inappropriate to its conceptualisation and one which, at the same time, relegated geography (in the form of space) to a passive explanatory role associated with the spatial distribution of the middle classes. In the time which we have been working on this research, two pieces of work have appeared which support our findings and enable us to account for them.

The first of these is Savage *et al.*'s (1992) study of the middle classes. As Savage *et al.* emphasise, one of the major problems with the radical human geography literature of the 1970s and 1980s was its obsession with capital as opposed to people, a tendency which is as applicable to contemporary geographical debate on the middle classes as it was to earlier debate on, for example, spatial divisions of labour (see Cooke, 1985). For Savage *et al.* this has led the geographical literature into the academic equivalent of a dead end *vis à vis* the middle classes. Here it is the density of the middle-class population in the south-east – explained in turn by the role of the south-east in the contemporary spatial division of labour – which attracts attention, to the detriment of everything else. But, as Savage *et al.* show, when the focus switches to the middle classes in contemporary Britain, and particularly to their spatial mobility, a very different pattern emerges.

The argument which Savage *et al*, put forward is one of the south-east as an 'escalator region' for the middle classes in contemporary Britain. As their work on the Longitudinal Survey shows, and contrary to the arguments of Massey (1988), the south-east is not a magnet region for the middle classes. Instead, it attracts large numbers of 'new entrants' from the educational system to middle-class forms of work; it promotes these recruits (and its own young people) at accelerated rates into positions of responsibility; and it exports a significant number of established middle-class people to other regions – at the time (1971–81), apparently to the south-west, East Anglia and the East Midlands (Savage *et al.*, 1992: 182). These findings suggest that the old distinctions between the provincial and London middle classes (founded on the regional economies of the nineteenth-century spatial division of labour) are breaking down:

> Rather than each region having its own distinct middle class, the processes of migration tend towards a circulation of people around the country, leading to a more homogeneous middle class. . . . Many regions will have a number of ex-Southerners in their middle classes [and] many of the children of middle class parents throughout Britain will spend time in London and the South east.
>
> (Savage *et al.*, 1992: 169, 184)

In such circumstances uniformity rather than difference is to be anticipated.

It was indeed, precisely the pattern of spatial mobility outlined by Savage *et al.* which we encountered within our two sets of employer interviews. Thus, in the north-east, we found ourselves interviewing only two households (out of a total of twenty-nine) comprising 'home bred' north-easterners. The remainder were immigrants, and included a sizeable number of 'ex-Southerners', many of whom had come to Newcastle or Durham straight from London and/or the south-east region. In the south-east, of the forty couples interviewed, only two had been born and brought up in the south-east. The remainder were 'new recruits' to the south-east middle classes and, for the most part, had moved into the region from higher education. Essentially then, the people whom we interviewed as employers of waged domestics were, and are, part of a nationally homogeneous, rather than regionally specific, middle-class. It is this degree of homogeneity amongst our employer respondents – their spatial mobility, as opposed to a rootedness in place – which we consider to have been critical in producing at least part of the uniformity which we uncovered in our work.

Whilst, homogeneity is a central facet of Savage *et al.*'s argument, so too is the notion of a fragmented middle-class, that is, a middle class comprising professional, managerial and petit bourgeois elements. Such arguments, together with those of spatial mobility, suggest that the differences in waged domestic labour – if they exist at all in contemporary Britain – might be ones manifested within local areas, rather than between them. Perhaps the differences would be between the professionals of Gosforth, Jesmond and Caversham on the one hand and, on the other, the petit bourgeoisie and managers living in the same areas? Perhaps in thinking initially about differences between our two study areas we were looking in the wrong direction? Our work provided no evidence to support such an argument, although certain political differences between these fragments of the middle classes, both generally and specifically (with respect to the politics of childcare provision), were revealed (and see too, Barlow and Savage, 1986).[17]

The picture which our work suggests then, is a uniformity which spans major lines of division within the contemporary middle class. As we show in Chapters Four, Six and Seven, this uniformity has a very great deal to do with the interweaving of dual-career household coping strategies with respect to social reproduction (and its organisation) with ideas which operate at the national level concerning both childcare and middle-class culture. In what follows, therefore, we demonstrate that the space which matters in the analysis of waged domestic labour is that associated with the middle-class household and with national level ideological debate. Moreover, we show too that the decision to employ waged domestic labour, for many of the middle-class households which we interviewed, was a very long way from being an assumed, automatic and accepted social practice.

It is, of course, precisely this type of argument which Lydia Morris puts forward in her engagement with the locality debate (1991). Thus, Morris argues that household arrangements are not necessarily best understood in relation to local labour markets and their associated social practices. Instead, in Morris's work on working-class households with unemployed male partners in both South Wales and Hartlepool, it is consistently factors internal to the household (particularly households' social networks) and those which operate at the national level (for example, the structure of the benefit system) which have proved important to understanding the organisation of domestic work and the form of the domestic division of labour. On the basis of our findings, we would support strongly the general form of Morris's arguments. There is no inevitability that local labour markets and/or localities will matter in the analysis of all phenomena. Instead, such causal potential needs to be demonstrated. And in the case of waged domestic labour in contemporary Britain, we have found local areas to be of minimal explanatory importance.

Whilst space in the form of local areas and their associated social practices may not occupy a pivotal role in our conceptualisation of the

Figure 2.18(a) Sources of outside-region advertised demand, 1981–91 (Great Britain)
Source: *Newcastle Evening Chronicle*

resurgence of waged domestic labour amongst the middle classes of contemporary Britain, in the form of patterns and distributions it certainly has had some influence on the matching of labour supply with demand. As we now show, this is the case for both categories of waged domestic labour investigated here, but for different reasons.

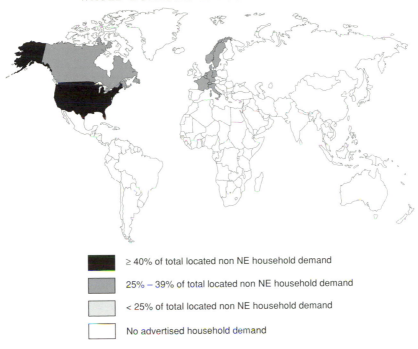

■ ≥ 40% of total located non NE household demand

▨ 25% – 39% of total located non NE household demand

□ < 25% of total located non NE household demand

□ No advertised household demand

Figure 2.18(b) Sources of outside-region advertised demand, 1981–91
(International)
Source: Newcastle Evening Chronicle

As previous sections in this chapter have shown, the distribution of the middle classes in Britain, and the pattern of advertised demand for the nanny form of waged domestic labour, were concentrated in London and the south-east region through the 1980s. As might be anticipated there is clear evidence to suggest that this demand for nannies was met not just by labour from within the south-east region, but by 'live-in' female labour from the provinces. Thus, for example, nine of the twenty-five nannies interviewed by us in the north-east had at some point during the 1980s worked in this capacity in London and/or the south-east region, before returning to work as nannies in the north-east. Moreover, within our north-east advertisement data, a total of 257 advertisements (20 per cent of the total recorded) – most of them for nannies – originated from non-north-east households, with a further 77 (6 per cent) coming from non-north-east agencies.

The major source of demand for labour from outside the region was London and the south-east. Indeed, London itself accounted for 41 per cent of these advertisements, with the by now familiar counties of Surrey, Berkshire, Kent, Hertfordshire, West Sussex (and also Essex) adding a further 15 per cent (Figure 2.18a). Throughout the 1980s then,

47

a considerable number of middle-class households living in London and/ or the south-east were seeing the north-east as a potential source region for the nanny and/or nanny/mother's help forms of waged domestic labour. Furthermore, a significant number of advertisements (most of these for au pairs, rather than for nannies) originated from households located outside the UK. Such households accounted for over 27 per cent of the non-north-east household total and, for the most part, stemmed from the United States (Figure 2.18b).

Whilst the presence of significant numbers of advertisements from non-north-east households certainly attests to the perceived role of the north-east as a source of childcare-related waged domestic labour, the rise of this form of employment in the north-east itself through the 1980s would appear to have been influencing the availability of young female migrant waged domestic labour within the region. Indeed, none of the nannies whom we interviewed in the north-east were considering moving out of the region to work. Moreover only 13 per cent of those young women whom we surveyed who were then on a childcare training course in the north-east were contemplating seeking employment out- side the region.[18] Whilst it is unrealistic to extrapolate beyond our study areas, such findings would seem to suggest that dual-career middle-class households in London and/or the south-east may find it increasingly difficult in future to satisfy their demand for nannies and or nannies/ mothers' helps by employing white female labour from the provincial regions. In such circumstances, and as we show in Chapter Three, there are plenty of lessons from a variety of different historical and spatial contexts which show how demand in London and the south-east could, in future, be met.

The foregoing shows how space, in the form of the spatial location and distribution of the middle classes in contemporary Britain, has influenced patterns of demand for the nannying form of waged domestic labour through the 1980s and how the shortfall in this particular category of labour in London and the south-east has conventionally been met. Finally we consider the role of space and place in relation to labour supply and demand with respect to the cleaner category of waged domestic labour.

Whilst advertisement data at both the national and local scales did not reveal the cleaner to be a major form of waged domestic labour in contemporary Britain, our survey work certainly did. The pattern of demand for cleaners revealed by advertisement data then is a serious underestimate of total demand for this form of waged domestic labour. One of the problems with this under-representation is that it is impossible for us to produce maps of national or local demand for cleaners. However, we are in no doubt that there are some quite significant differences between our two areas concerning labour supply. As we show in Chapter Five, whilst there were no differences between our two case study areas in terms of the social characteristics of the cleaning labour force, these same social characteristics meant that middle-class households in the two areas experienced different levels of labour availability. In the north-east, the middle-class households interviewed reported no difficulty in finding the cleaner form of waged domestic labour. Indeed, a number of our respondents had interviewed several people before employing a cleaner. In contrast, many of the households whom we interviewed in the south-east experienced considerable difficulties in finding this form of labour.[19] We consider this situation to be indicative of the contrasting levels of 'unemployed' and/or 'benefit-dependent' households in the two study areas throughout the 1980s, for – as we shall see in Chapter Five – it is these households which constitute one of the major sources of cleaning labour in contemporary Britain.

CONCLUSIONS

This chapter has established the extent of the resurgence in demand for waged domestic labour in Britain through the 1980s and its strong association with the care of pre-school-age children and general household cleaning. We have also shown this resurgence to exhibit a distinctive geography; one which manifests the spatial distribution of the middle classes, both nationally and within specific urban areas, in this case in London, Newcastle upon Tyne and Reading. However, we have emphasised that the geographies of advertised demand for particular categories of waged domestic labour in specific local areas are not necessarily identical. Indeed, the precise nature of these geographies and their potential for coincidence are shaped by the operation of specific middle-class housing markets. Thus, in Newcastle upon Tyne – a city with long established and strongly demarcated middle-class housing areas – it is hardly surprising that the spatial distribution of middle-class households advertising for nannies coincides with composite maps of advertised demand for waged domestic labour. Nevertheless, as much research conducted in the 1980s on housing generally and on gentrification specifically showed, other British cities have revealed major shifts in the geography of the middle classes. This has long been recognised to have occurred in London but it would also appear to have been the case in the Reading area, where there was also a distinctive shift in the geography of advertised demand for nannies through the 1980s. Notwithstanding the existence of these distinctive geographies, we have argued that geography (in the form of place-specific local social practices) is of minimal importance to the conceptualisation of waged domestic labour in contemporary Britain. This we have related to the arguments of Savage *et al.* (1992) regarding the spatial mobility of the middle classes, as well as to Morris's arguments regarding the household and its conceptual relationship to the local labour market (Morris, 1991). In the form of space – as we have stressed – we consider geography to be important to understanding the relation between demand and labour supply in particular local labour markets.

In addition to establishing the extent of the resurgence in demand for waged domestic labour in Britain in the 1980s and its attendant geography, the chapter has also considered the incidence of waged domestic labour within middle-class households. On the basis of our survey findings, we suggest that between 30 and 40 per cent of dual-career households in contemporary Britain employ waged domestic labour in some form; that around 40 per cent of such households with pre-school-age children employ a nanny; and that three-quarters of those dual-career households employing waged domestic labour employ a cleaner. Moreover, as our findings show, a minority of dual-career households (less than 15 per cent) employ more than one waged domestic (typically a cleaner and a nanny). Waged domestic labour therefore is revealed to be a vital facet of daily social reproduction amongst the middle classes of contemporary Britain.

In the remainder of this section of the book we focus on accounting for the resurgence in demand for waged domestic labour amongst the middle classes of contemporary Britain.

3

PERSPECTIVES ON WAGED DOMESTIC LABOUR

In this chapter we consider the existing literature on waged domestic labour, and establish the contours of the theoretical position developed in Chapters Four and Five. The chapter is divided into two main sections. Firstly, we review the literature on waged domestic labour, focusing on waged domestic work in a variety of historical and contemporary settings and on the 'dual-career family'. Our aim is not to provide an exhaustive summary of these fields. Rather, we identify general trends, the intention being to establish the historical and geographical extensiveness of waged domestic labour, its commonalities and particularities. This underpins the second section of the chapter, in which we review and critique the various theoretical perspectives on waged domestic labour. This serves to prefigure some of the features of our conceptualisation of the resurgence in demand for waged domestic labour in Britain in the late 1980s.

WAGED DOMESTIC LABOUR: AN OVERVIEW

Waged domestic labour in historical context

By the last quarter of the nineteenth century, a vast array of maids, nurses, cooks, coachmen, gardeners, laundresses and charwomen served the personal needs [of] . . . the upper and middle ranks of society. Everybody who was anybody kept a maid.

(Ebery and Preston, 1976: 2)

The above constitutes a familiar picture, one which has become part of the collective image of Victorian and Edwardian Britain through the popular culture of television dramatisations such as *Upstairs, Downstairs, You Rang, M'Lord?* and *Jeeves and Wooster*; through fiction, from *Jane Eyre* and the novels of Ivy Compton Burnett to the works of Margaret Forster and Catherine Cookson; and through the increasingly visible histories of working-class British women (see, for example, Harrison, 1975;

51

McCrindle and Rowbotham, 1977; Newton, Ryan and Walkowitz, 1983).[1] That this is so, is unsurprising. In the nineteenth century, and up until 1914, domestic service comprised one of the most important occupational categories in Britain (Burnett, 1974). Nevertheless, it is only relatively recently that it has become a topic for academic investigation (Ebery and Preston, 1976, and Higgs, 1983, 1986a, 1986b).

Ebery and Preston categorise employment within domestic service in Britain into three distinct phases: the expansion years (1851–71); the period 1881–1901, characterised by moderate decline; and 1901–51, a phase of rapid contraction. Domestic service occupied over 1 million individuals at least throughout the period 1851–1907, but expanded dramatically through the second half of the nineteenth century, during which time around 2 million were thus employed. Even in the inter-war period employment within domestic service did not contract at anything like the rate which is often assumed. Indeed, as late as 1951, nearly 1.8 million were thus occupied, 1.3 million of these being women (Taylor, 1979). This, then, was overwhelmingly a female occupation, and between 1851 and 1901 it accounted for over 40 per cent of all employed women (compared to between 3 and 5 per cent of all men in employment). Not surprisingly, the ways in which men and women were employed in domestic service varied. 'Indoor service' was provided predominantly by women, with over 80 per cent of all women employed within domestic service being thus occupied, with a minority working as 'auxiliaries', particularly as charwomen and laundresses. Very few women worked as 'outdoor servants' (compared to around 50 per cent of males in service). Furthermore, for women in domestic service, strong divisions along the lines of age were apparent. Hence, 'indoor service' was predominantly a young woman's occupation. In 1911, for example, 54 per cent of female indoor servants were aged between 15 and 24, with only 11 per cent aged 45 and over. In contrast, 'charring' was the preserve of the older, married or widowed woman.

As well as documenting numbers employed in domestic service through the Victorian and Edwardian periods, Ebery and Preston's work maps the geography of domestic service. The distribution revealed is marked and uneven. Interestingly, too, it resonates strongly with the spatial distributions revealed in Chapter Two. Working on the basis of the number of servants per head of population, Ebery and Preston argue that areas of high and low servant concentration can be identified. Thus industrial and/or mining areas (such as South Wales, the East and West Midlands, Lancashire, Yorkshire and the north-east) are shown to have been areas of low servant concentration, whilst areas of high concentration occurred predominantly in southern Britain (although pockets of the provincial cities also fell into this category).[2] The highest figures of domestic servant concentration, however, occurred in certain areas of London and the south-east, with Hampstead, Harrow, Ealing, Chelsea and Westminster recording particularly high figures.

In a similar documentary vein to Ebery and Preston's research is that of the social historian Edward Higgs (1983, 1986a, 1986b). Higgs's primary concern with respect to domestic service in Britain has been to emphasise the difficulties with many of the sources used by historians of domestic service. Singled out for particularly strong criticism are those accounts which rely on manuals of domestic service, and which stress, consequently, the patterns of domestic service employment within the homes of the aristocracy (see, for example, Horn, 1975). Instead, like Ebery and Preston, Higgs prefers to rely on the Census Enumerator's Books. None the less, he emphasises the considerable difficulties which surrounded the use of the term 'servant' by census enumerators in the nineteenth century and estimates that varying recording practices could have over-estimated the absolute number of domestic servants in Rochdale by as much as 33 per cent; a figure which, if replicated nationwide, has massive implications for the numbers considered to have been employed within domestic service in nineteenth-century Britain.

Beyond a concern with absolute numbers and geographical patterns, the literature on nineteenth-century domestic service in Britain has focused on the nature of work within this occupational category. And, once again, it is the sources used which have proved critical to representations of domestic service work. For those who have relied on manuals of domestic service, the picture conveyed is of a hierarchy of domestic service, in which gender and the control of others figures significantly (Banks, 1954; Burnett, 1974; Hecht, 1956; Horn, 1975; Huggett, 1977; Marshall, 1949). In contrast, authors who have worked from the Census Enumerator's Books have emphasised the numerically more important, but largely invisible, 'maid of all work'. Their intention has been to counter the populist 'Upstairs Downstairs' conception of Victorian domestic service. For most people who employed household domestic workers in Victorian and Edwardian Britain, a young, un-married female 'maid of all work' from a rural area was the norm, and not veritable armies of liveried and/or uniformed staff (Higgs, 1986a). Such an emphasis has led Higgs to consider nineteenth-century domestic service in Britain as a life-cycle occupation, and specifically as a job for young women before marriage, particularly in areas with no alternative forms of female employment.

The other main strand of debate within the historical literature on domestic service in Britain is the relationship between employers and servants. Part of this discussion has been conducted at the level of changes to the master–servant relation (Hecht, 1956; Macpherson, 1973). However, running parallel to this legal discourse has been a broader debate over the status implications of servanthood and servant-employing. This type of analysis is encapsulated in Banks's portrayal of the servant as part of 'the paraphernalia of gentility', and by Booth's use of the presence of servant(s) as diagnostic of the Victorian middle-class (Banks, 1954; Booth, 1903).

Domestic service in Victorian and Edwardian Britain has also received attention from feminist historians (Branca, 1975; Davidoff, 1973a, 1974, 1983; Jamieson, 1990; Vicinus, 1972). At one level this work can be read as 'making women visible' (Rowbotham, 1977). At another, it shows how the social and status distinctions present within society are exhibited in reproductive space. It is in Davidoff's work that such arguments are most clearly articulated. Here, the reproductive space of the Victorian middle-class household is shown to be socially and spatially segregated. Servants are shown to have been confined to certain social spaces; to have been constrained in their use of other spaces, but, at the same time, to have been used by employers to control the access of non-household members to the protected space of the Victorian middle-class household. Such actions served to signify the place of servants in the social order but they also played a vital part in the construction of middle-class and working-class versions of femininity.

Whilst the vast majority of historical work on domestic service pertains to Britain, a considerable amount of research has been conducted on nineteenth-century domestic service within the United States.[3] Following Birch (1984), much of the early literature in this field can be divided into two categories; a collection of work on 'the servant crisis' at the turn of the century (Eaton, 1899; Laughlin, 1901; Pettengill, 1903; M. Robinson 1924; Salmon, 1897); and a series of general surveys of domestic service in this period, based on the accounts of Salmon and others (Dudden, 1983; Sutherland, 1981). Katzman's (1978) work is typical of the latter and focuses on three issues: the changing composition of the servant class in different areas of the United States over time; the changing form of employment in domestic service; and the nature of the work. Nineteenth- and early twentieth-century domestic service in the United States is shown to differ little from its counterpart in Britain. Long working hours, heavy physical labour and social isolation were the norm, along with a high degree of job mobility. Katzman also points to the change from live-in to live-out domestics by the 1920s, a trend which he links to the changing labour force composition of domestic service. Nineteenth-century domestic service in the southern states is shown to have been the domain of blacks. In contrast, in the northern and eastern states, immigrant white women, particularly Irish women, comprised the majority of the domestic service workforce. By the 1920s, with the expansion in alternative forms of female employment and the migration of black women north from the southern states,

United States domestic service had become a predominantly black female occupation. Indeed, Katzman estimates that at this time domestic service occupied 75 per cent of black female wage earners.

A very different emphasis is provided by Bettina Birch (1984). Her argument is one which focuses primarily on the flaws in Salmon's representation of the United States servant crisis. Far from being the relatively high-waged occupation, which Salmon maintained, Birch establishes that domestic service was poorly paid relative to other occupations, and that it was this, rather than stigma, which was behind the servant crisis in the United States. Beyond this, Birch suggests that the decline in the number of servants was itself ultimately responsible for the 'technological fix' with respect to housework. She interprets the servant crisis in the United States in the early twentieth century as resulting in the transformation of housework; initially, from low-waged work to non-waged work performed by 'housewives' and then to unpaid work performed by women in conjunction with a range of technological appliances (Birch, 1984: 199). A similar concern with moving beyond broad-brush, overly simplistic and frequently contestable generalisations on nineteenth-century domestic service in the United States underpins the contributions to a recent edition of *Social Science History* (see Cole, 1991; Gross, 1991; Lintelman, 1991).

Waged domestic labour today

Many of the above strands of debate are also discernible in the literature on contemporary waged domestic labour, where the nature and characteristics of waged domestic work and domestic workers, the question of occupational mobility, employer–employee relations, questions of status, domestic workers' struggles to organise politically, and the question of the persistence of waged domestic labour are all central to debate (see, for example, Bunster and Chaney, 1985; Chaney and Garcia Castro, 1989; Gill, 1988; Gogna, 1989; Jelin, 1977; Nett, 1966; Radcliffe, 1990; Rubbo and Taussig, 1978; Smith, 1973, 1989; and Young, 1987 – all on waged domestic labour in Latin America). Many of the same themes also transfer to studies of waged domestic labour in contemporary South Africa. Here the definitive study is that by Jacklyn Cock (1980a). Working from a survey of 225 waged domestic workers, most of them middle-aged and living in the Eastern Cape, Cock documents graphically the collective and individual work experiences of these women. Beginning with an account of the conditions of employment, the low wages, excessively long hours, hard physical labour and live-out nature of the work, Cock builds up a picture of the South African waged domestic as a 'trapped worker'. Beyond this, much of Cock's concern is with the relationship between waged domestic workers in South Africa and their employers. This she represents as a 'paternalistic' form of dependence, in which the domestic worker, whilst being seen as 'part of the family' is considered and treated as a child.

A similar concern for the relationship between employers and their domestic workers features in Eleanor Preston Whyte's research (1976). Using two contrasting neighbourhoods in Durban, one prosperous and middle-class, the other culturally mixed and characterised by a range of income groups, Preston Whyte shows how the employer–employee relationship is a long way from being as homogeneous as Cock suggests. In 'Ridgeheights' (an area characterised by a gender division of labour in which the man's role is that of the successful businessman/professional employee and where the woman's entails a round of social and civic responsibilities) the relationship is one of formality and distance. Modelled on the populist conceptualisation of domestic service in Victorian Britain, the 'Ridgeheights' domestic workers had set responsibilities and routines, and were spatially as well as socially separated from the worlds of their employers. In contrast, in the suburban 'family' homes of 'Central Flats', relations of close physical proximity and familiarity prevailed, with domestic work performed both by waged domestic workers and female household members.

Shifting the geographic focus once more, the literature on contemporary waged domestic labour in the United States resonates strongly with that on South Africa. Again, the dominant research questions are those of who does this form of domestic work and why, and the nature and conditions of the work and relations between employers and employees. There are, for example, several studies of minority women of colour working as household domestics (Colen, 1986; Glenn, 1980, 1981; Romero, 1988). All devote considerable space to discussions of waged domestic labour and occupational mobility, as well as to the work itself. The work of Romero (1988) is a case in point. Drawing heavily on Evelyn Glenn's work on Japanese-American women, Romero argues that minority Chicanas find themselves trapped in waged domestic work. Such arguments are similar to those of Cock in the South African context. As might be expected, Romero also considers in some detail the relationship between employers and their domestics. Notwithstanding the degree of job autonomy exhibited by Chicanas (particularly in terms of work scheduling), Romero argues that minority women represent cultural curiosities to their employers (Romero, 1988: 86), and that this is indicative of the racism which underpins the social relations of waged domestic labour in the United States. Finally the question of status is addressed; although from the viewpoint of waged domestic workers rather than employers. For Romero, the most important issue here is the social stigma attached to waged domestic labour, and the strategies which Chicanas use to counter this.

Similar in intent and concerns is Shellee Colen's study of West Indian childcare and domestic workers in New York City. As with Romero's Chicanas, Colen's respondents are migrant women and for Colen migration constitutes a vital aspect of any understanding of contemporary waged domestic labour in the United States. It is, therefore, to the nature of waged domestic work in the sponsorship situation which Colen directs attention. The picture presented is summarised as 'a form of legally sanctioned indentured servitude until the green card is granted' (p. 52). For these West Indian women, the sponsor job trapped them in forms of work characterised by exceptionally long hours, abysmally low pay, heavy physical labour and social isolation.

Colen's discussion of employer–employee relations emphasises the grounding of the personalised relations of waged domestic labour in divisions of class, gender and ethnicity. Waged domestic workers, she argues, are constructed as 'other'; a situation manifested in the various means of social demarcation used by employers, ranging from lack of respect through to the use of uniforms and including living and eating arrangements. Playing a central legitimating role here is the ideology of the family. Phrases such as 'she's one of the family' are suggested by Colen to ease the use of waged labour in the personalised context of the household; to coerce individuals into performing heavy physical tasks and to help them tolerate 'exploitative' working conditions.

Where Colen departs from Romero is over what happens to women who work as paid domestics. Romero's argument is essentially identical to that of Cock; waged domestic labour for Chicanas is a lifetime, and not a life-cycle, occupation. Colen, in contrast, maintains that once having gained their green card, West Indian women become upwardly occupationally mobile (Colen, 1986: 53). In the majority of cases, studied by Colen, this mobility involved a movement into either clerical or health service occupations.

At a more general level is Judith Rollins's study *Between Women*, based on her own experiences as a waged domestic worker (Rollins, 1985). Notwithstanding the difference in approach, the same themes dominate the discussion. Her description of the nature and conditions of waged domestic labour accords strongly with both the in-depth studies of minority women reviewed above and the survey work conducted by Cock in South Africa. The job is shown to be physically demanding, characterised by long hours, low pay, monotony and social isolation. Rollins also examines employers and domestics in more detail, and the relationship between them. Employers are identified as white, in the middle to upper income range, and female. All are seen to assume that the responsibility for household work is theirs and not men's, and four reasons for employing waged domestic help are suggested: practical necessity; to enable non-working hours to be used for more valued activities; to symbolise status; and to continue a family tradition of employing domestic help. Regarding waged domestic workers, Rollins agrees with those who have examined the experiences of particular groups of minority women working as domestics. The structuring of the labour market along class, gender and ethnic lines is seen to restrict the employment opportunities of women of colour, and with limited education, to waged domestic labour.

In line with many other writers on contemporary waged domestic labour, Rollins sees the relationship between employers and employees as exploitative. Like Cock, she emphasises the importance of deference, obligation, loyalty and respect within the relationship, and represents it as characterised by maternalism and imbued with childlike qualities. However, where Rollins departs from the tradition of much of the writing on contemporary waged domestic labour is in her willingness to consider this in terms of its implications for gender relations. In essence, Rollins argues that the employment of household domestic workers is reproductive of existing gender inequalities and of the gendering of domestic work as 'women's work'. Employers she sees as being '. . . willing to take full advantage of the class and racial inequalities generated by this social system to mitigate against their gender disadvantage' (Rollins, 1985: 185).

Regardless of geographic focus then, the literature on contemporary waged domestic labour is remarkably consistent in its empirical findings. Wherever it occurs this work is characterised by hard physical labour, long hours, low pay and, frequently, poor employment conditions. Divisions of class, gender and ethnicity mean that it is almost invariably women who work in this occupation, and that most of these women will be young migrant women of colour. To a degree there is some difference of opinion as to whether waged domestic labour currently constitutes either a life-cycle or a lifetime occupation but this would appear to be influenced by the existence of alternative employment opportunities for women in particular areas and by the availability of new categories of migrant women to work as household domestics. In South Africa, where black women remain 'at the bottom of the pile', waged domestic labour represents a lifetime occupation. In contrast, in the United States the in-migration of Hispanic and south-east Asian women has meant that for Colen's West Indian women, waged domestic labour may now be no more than a life-cycle occupation en route to the attainment of the green card. Finally, there exists considerable agreement as to the form of relations between employers and employees. Essentially relations between women, these relations are, it is argued, constructed on maternalist lines; the domestic is treated as, and expected to behave as, the quasi-child – deferential, subservient and obedient, in short 'as a maid'.

Having said this, the empirical literature on waged domestic labour also suggests that the form which waged domestic labour takes, and has taken, in different contexts is variable. Historically, whether it is men or women who do waged domestic labour, varies (Tranberg Hansen, 1986a; Gaitskell *et al.*, 1984). There are also contrasts at analytical and experiential levels. Thus, whilst the experiences of waged domestic labour in the nineteenth century of, say, black women in the southern states and young rural migrant women in London may have been similar, the precise mesh of social forces which combine to structure the social characteristics of paid domestics in these cases is subtly different. Gender may be a commonality here, but it has not always been so (Cole, 1991). Ethnicity figures, but in one of these cases it is black women and, in another, white women who are being constructed as waged domestics. Age (the life-cycle) enters the equation too, but in subtly different ways, as some of these women are barely adult and others considerably older. Likewise, whilst there are similarities in the contemporary period in the work of West Indian women in New York City, Filipinas working in diplomats' houses in London and black South African women working in the suburbs of Johannesburg, the combination and conjuncture of class, ethnicity, gender and age is variable. Differences too are visible in the social relations of paid domestic work. Whilst notions of 'service', 'being maid-ish', of deference, servility and status have figured centrally in the construction of the social relations of waged domestic labour in different times and places, they cannot in any sense be seen as universals. Moreover, in other more contemporary circumstances, such notions have been overtly challenged. We consider the historical and spatial variation in form of waged domestic labour, to be extremely important theoretically. Thus, whilst we acknowledge the possibility of making theoretical generalisations about the phenomenon of waged domestic labour, it is particular forms in particular temporal and spatial contexts which we emphasise and which we take to be the key explanatory issue.

WAGED DOMESTIC LABOUR AND THE 'DUAL-CAREER FAMILY'

The second area of literature in which waged domestic labour figures relates to the 'dual-career family'. The term dual-career family first came to prominence in the late 1960s and is closely associated with the work of Rapoport and Rapoport (1969, 1971, 1976, 1978, 1980). Following on from their study with Fogarty on women and careers (Fogarty, Rapoport and Rapoport, 1971), the Rapoports argued that, whilst still a statistical rarity, the dual-career family was both a pioneering new family form and one which, in the light of greater female participation at higher levels in the labour force, was going to grow in significance (Rapoport and Rapoport, 1969). There emerged therefore in the late 1960s and early 1970s, a series of what Rapoport and Rapoport (1980: 1) later referred to as 'pivotal studies' of this 'new' family form (Epstein, 1971; Garland, 1972, and Poloma and Garland, 1971; Holmstrom, 1972; Poloma, 1972; Rapoport and Rapoport, 1971, 1976). It is worth highlighting the primary concerns of this early literature on the dual-career family and that which followed it, as well as the perspectives which informed it. These are important for understanding how much of this literature engages with the phenomenon of waged domestic labour.

For the Rapoports, the dual-career family is one in which 'both the husband and the wife pursue active careers and family lives' (Rapoport and Rapoport, 1971: 7). Both partners in the household are therefore involved in career structured occupations and play an important role in family life. The dual-career family then is distinguished from other family forms, notably the dual-worker (ie., dual-earner) family (in which both partners are in gainful employment, although not in career structured occupations), the 'two-person career couple' (in which the wife provides the emotional and social support structure to sustain her husband's career) and the then 'conventional' family form, comprising the male breadwinner and the housewife.

The Rapoports' analysis is one which at one level can be seen to reflect a trend then emerging in the family studies literature, namely to stress the importance of the interconnections between paid work and home (or family) life (Young and Willmott, 1973). At another level it is permeated by many of the dominant sociological concepts of the time, notably by social roles, norms and conventions – in this case, in relation to the family – and by individual attitudes to these. Such concepts reflect clearly the influence of Parsons, but it is to Goode and his role strain theory that the Rapoports looked for their principal conceptual direction (Goode, 1960). Thus, it is the stresses and strains encountered by the dual-career family which provide the central interest for the Rapoports throughout their research on this particular family form. Stress and strain are here conceptualised in terms of dilemmas, conflicts, barriers, problems and 'tension lines'. They therefore are considered predominantly at the level of individual men and women within dual-career households.

Subsequent research on the dual-career family has been cast in a similar vein. The studies by Bailyn (1970) and Bebbington (1973), for example, are concerned with marital satisfaction and with eliciting reasons for how and why particular individuals come to form the 'stressful' dual-career family pattern, when they are not compelled to do so. In contrast, research published through the late 1970s and early

1980s focuses on strain and strain management within the dual-career family (Rapoport and Rapoport, 1980: 34). Examples of such strains include; the rigidity of the occupational system (Bailyn, 1978; Rose, Jerdee and Prestwich, 1975); the combination of domestic and career roles (Johnson and Johnson, 1980); as well as the problem of geographical mobility (Duncan and Perucci, 1976; Mortimer, Hall and Hill 1976; Wallston *et al.*, 1978). However, strain management is considered predominantly in terms of the strategies and mechanisms proposed by Goode (1960). Thus, Johnson and Johnson (1980: 157–9) suggest a variety of intra-family strategies to ease the strains encountered by dual-career couples, although they also press for the implementation of a variety of social policy developments. Role cycling (the prioritising of male and female careers at different times) is advocated, alongside a series of more broad based changes rooted in the liberal feminist politics of equality of opportunity; for example, the removal of barriers to women's advancement; adequate, inexpensive childcare; career allowances for women with children and a change in the normative expectations of women.

This brief survey highlights the centrality of stress, strain, strain management and role strain theory within the literature on the dual-career family. It also emphasises as the principal area of research interest the way in which the dual-career couple functions as a partnership. Both points are important to understanding how waged domestic labour figures within this literature.

Perhaps the most important point to establish at the outset here is that waged domestic labour (or 'hired domestic help', as it is referred to in this literature) is a marginal concern for these researchers. Nowhere in either the work of the Rapoports, or in the work of the individuals cited above, does waged domestic labour merit more than passing, cursory attention. Typical in this respect are the following comments

> The most usual situation amongst our couples where both husband and wife pursue a career is a re-arrangement of the domestic side of their lives. *Some of the household tasks are delegated to others*, part is re-apportioned between husband and wife and children. In effect, each member pursues a career and performs some household and childrearing activities.
>
> (Rapoport and Rapoport, 1969: 9, our emphasis)

> *Surrogates were used in all cases and the wives directed them*. If the surrogates were kin, without exception they were the wife's kin – her mother, or in one case, a maiden aunt – and the management of the household was then shared by the women in a rather traditional way.
>
> (Epstein, 1971: 559, our emphasis).

61

There is, however, no correlation between the space devoted to waged domestic labour in this literature and its importance to individual dual-career families. Indeed, it turns out that waged domestic labour had been utilised extensively by almost all of the dual-career families investigated by these researchers. For example, in the case of the Rapoports' five original family case studies we find the following pattern. The Kileys (a science manager and a sales manager) 'relied on a series of German au pair girls employed through the Labour Exchange' (Rapoport and Rapoport, 1971: 35); the Bensons (both architects) employed a 'daily' and a 'live-in student' (p. 84); and the Jarretts (a TV drama director and an architect) were 'assisted by nannies . . . with a succession of au pairs and dailies later' (p. 138). Indeed, 'the Jarretts [even] . . . constructed a self-contained flat adjacent to their house in which domestic help . . . [was] accommodated' (p. 138). In all of the cases explored by the Rapoports, then, the employment of waged domestic labour is treated as a normal, necessary and unexceptional part of the constitution of the dual-career family pattern and its reproduction.

Both the normalisation of waged domestic labour within the dual-career family and the lack of comment which waged domestic labour and its use receives within the work of the Rapoports and the research which their studies inspired, can best be understood in terms of its conceptual location within role strain theory and its emphasis on the couple as the focus of research. Given the importance of role stress and strain within the Rapoports' initial formulations, domestic work came to be seen as an inevitable source and site of stress within the dual-career family, and as one of the principal generators of 'overload dilemmas' for couples. As a consequence, the vast array of waged domestic labour which the Rapoports refer to in passing (live-in, live-out, short-term, long-term, full-time, part-time, nannies, au pairs, dailies, students, secretaries and domestic help couples – Rapoport and Rapoport, 1969: 10) come to be seen as just one means of reducing couples' work overload. Waged domestic labour therefore is considered a mechanism for stress management and a signifier of role strain within the dual-career family. Given this, waged domestic labour can only be presented as one amongst a number of 'couple strategies' for stress/strain alleviation, alongside many other, mostly socio-psychological options, for example, positive thinking, value clarification, compartmentalisation, compromise, tension line articulation and personal negotiation (Lawe and Lawe, 1980).

Whilst the Rapoports were certainly criticised for their 'assumption that the pattern [of dual-career families] is made possible only through the use of elite resources such as servants' (1976: 12), they were not criticised for their conceptual approach to waged domestic labour within the dual-career family. In responding to their critics, the Rapoports attempted to counter this elite tag. Thus they showed that many of the members of dual-career partnerships were from working-class backgrounds and had not in fact used waged domestic labour until they became committed to the dual-career family pattern (ibid.). In addition, they argued that 'conventional families' also use support workers (in the form of female kin) and that 'domestic work can be a job like any other and under appropriate conditions need not be exploitative' (ibid.: 12–13). Interestingly, waged domestic labour did not figure on their research agenda for the 1980s. However, by the late 1970s and early 1980s, the centrality of waged domestic labour to the dual-career family form was beginning to be acknowledged by other researchers.

In their 1977 paper, Hunt and Hunt, whilst concerned mainly with the relation of the dual-career family to historical and potentially emergent family forms, remarked:

> Hired help on a single family basis involves a category of workers that must be paid out of the take home earnings of the nuclear unit. Consequently, the dual-career family is premised upon the increased use of a class of workers locked into a standard of living considerably lower than their employers . . . it would provide for the 'liberation' of one class of women by the continued subjugation of another.
>
> (Hunt and Hunt, 1977: 413)

This is clearly a step beyond criticisms on the grounds of the use of elite resources. Instead, this is the first explicit recognition within the dual-career family literature that gender equality for some women may rest upon the perpetuation of class differences between women. Such themes have been pursued further in the most recent generation of work on the dual-career household, notably by Rosanna Hertz (1986).

Hertz's work represents some of the most recent research in the dual-career family tradition and, whilst pursuing similar themes to those investigated by the Rapoports fifteen years earlier, contains some important differences. The most obvious of these is her focus on dual-career couples in the corporate world. The corporate world, it is argued, is not just an increasingly important source of employment for the dual-career family, but poses a rather different set of potential problems for the dual-career family form than the types of employment which figured in the earlier generation of dual-career family studies. Notable in this respect is the effect of corporate cultures, as well as the problems which dual-career couples face in trying to combine the needs of the household with those of two, usually spatially extensive, corporations. For our purposes, though, it is two further differences which are of greater importance. The first of these is Hertz's explicit concern with the establishment of gender equality. The second is her conceptualisation of the dual-career family. For Hertz, the major research question to which her dual-career family study is related, is 'what will it take for men and women to be truly equal?' (1986: xi). Economic equality is her answer; a situation which, she maintains, will provide the basis on which to establish social equality. The dual-career couple, Hertz argues, represents a form of household in which men and women are not just economic equals but social equals too. Furthermore, relative economic equality is said to enable a major shift in traditional gender roles, such that it is impossible for the notions of breadwinner and homemaker to be applied to (or accord with) either partner in a dual-career couple. However, the question which Hertz goes on to pose is whether equality for the dual-career couple is based on 'the availability of other people ... for whom equality is not common' (p. ii). Her answer is an unqualified 'yes', and it is this which leads her to conceptualise the dual-career couple in terms of a reciprocal relationship between autonomy and contingency (p. 196). Whilst autonomy is generated by the financial and life-style independence associated with the dual-career set up, Hertz maintains that the existence of the dual-career family is contingent upon three things: the availability of career positions; the availability of outside labour to perform household work and childcare; and the continued capacity of couples to adapt to competing employer demands (Hertz, 1986: 196). Unlike the Rapoports, and the studies inspired by their work, Hertz, like Hunt and Hunt, recognises explicitly the centrality of waged domestic labour to the reproduction of the dual-career family pattern:

> The couple's capacity to sustain two careers (and to acquire them in the first place) often depends on the availability of someone else to perform, for a wage, the duties necessary to maintain the household.
>
> (Hertz, 1986: 6).

In addition, she moves beyond observational patterns to root her explanation for the use of waged domestic labour in the occupational class structure. Rather than seeing dual-career families as built upon the direct exploitation of working-class couples or the working-class wives who work as household domestics, Hertz argues that the occupational class structure produces two groups in relation to household work and childcare: 'those who need the work done but who do not have the time to do it, and those who have little alternative but to do it for pay' (Hertz, 1986: 202). Effectively, the employment of paid domestic help by dual-career couples is considered to reflect the occupational class structure (ibid: 203).

Somewhat surprisingly, the debate engendered by Hertz's work on the dual-career family has neither examined nor developed her conceptualisation of the dual-career family. Instead, debate continues to emphasise the conceptual importance of role strain and minimise the importance of waged domestic labour. Thus, and contrary to Hertz, Moen and Dempster McClain (1987) argue that professional women in dual-career households use their financial resources to reduce their work involvement, particularly when their children are young. This is seen as a mechanism to reduce role strain and role conflict. Indeed, reduced female labour force participation is seen as the major dual-career couple coping strategy by Moen and Dempster McClain. Similar arguments are made by Grant *et al.* (1990: 43) on the basis of a survey of 204 male and female physicians, despite the fact that 4 per cent of this sample employed full-time (i.e., five days a week) domestic help and 35 per cent part-time help, typically on a one-day-a-week basis.

Whilst the literature on the dual-career family, then, has certainly touched on waged domestic labour, for the most part the latter has been very much 'backstage'. In the vast majority of studies of the dual-career family, waged domestic labour figures as an assumed and unexceptional given in the reproduction of this particular household form. It is conceptualised as one amongst a number of means by which partners can cope with the stresses and strains of domestic work, parenthood and two full-time careers. Only in the work of Hertz (1986) and the Hunts (1977, 1982) do we find acknowledgement that the reproduction of the dual-career family pattern is both dependent on the existence of female domestics and, in turn, generative of major, class-based inequalities between women.

CONCEPTUALISING WAGED DOMESTIC LABOUR

In comparison to the volumes of material written on waged domestic labour in different historical and geographical contexts, little space has been given over to the overt conceptualisation of this form of work. Moreover, with the one notable exception of Hertz's work, the literature on the dual-career family has had little enough to say about waged domestic labour at the empirical level, let alone about its conceptualisation. In what follows, the various overt and covert conceptualisations of waged domestic labour present within these sets of literature are considered and evaluated with respect to how they would inform the analysis of the resurgence of waged domestic labour within contemporary Britain.

Waged domestic labour within the modernisation thesis

Whilst the Third World studies and development literature written from a modernisation perspective is considerable, its authors devote little

65

space to waged domestic labour (see, for example, Rostow, 1960). There have, however, been a few attempts to conceptualise waged domestic labour using the modernisation framework, as well as many empirical accounts which utilise facets of this argument. The most explicit attempt to situate domestic service within the modernisation thesis is that of Coser (1973). Coser's central argument is that the role of the servant is obsolete in a modern society. This, he maintains, reflects the increasing unacceptability of an occupation based on status, strong household ties, non-specific obligations and an uncertain line of division between home and work. In short, Coser argues that domestic service is a 'pre-modern' occupation, whose characteristics of personal allegiance, long hours and high levels of commitment were legitimated by religion and enforced by an economic context which offered no alternative employment opportunities. The demise of religious legitimation and the opening up of new forms of employment are seen to spell the end for domestic service. For Coser, domestic service represents something which only 'an underclass of social inferiors who have no place in the social scheme of things' will currently accept (p. 39).

Coser's ideas have found a degree of acceptance within the historical literature on domestic service, most notably from writers producing overview summaries. A case in point is Chaplin (1978: 123), who accepts both the argument which links the decline of domestic service to industrialisation/modernisation, and the importance of alternative employment opportunities to the reduction in the numbers employed in service (and see too Fairchilds, 1979). More generally, though, it is the availability of alternative employment opportunities, rather than any overt reference to the modernisation thesis itself, to which most authors refer. For many writers on domestic service, the presence/absence of employment opportunities elsewhere represents the reason for the decline/continuance of domestic service. Burnett (1974), Ebery and Preston (1976), Higgs (1983, 1986a) and Horn (1975), for example, all regard the expansion of manufacturing activity as critical to the decline of domestic service in Britain. Likewise, Katzman (1978) accounts for the changing labour force characteristics of domestic service in the United States in precisely the same terms. Although their acceptance of the modernisation thesis remains largely implicit, such interpretations show the extent to which modernisation pervades much of the writing on domestic service.

Also located within the modernisation perspective are those authors who conceptualise domestic service as a 'bridging occupation'. The issue here is not domestic service's various 'pre-modern' characteristics but the role which domestic service plays in assimilating individuals within a new and/or different society. The clearest statement of domestic service as a bridging occupation is provided by Broom and Smith (1963).

They maintain that grooms had the potential to become stable owners; wine stewards, wine merchants; cooks, restaurateurs; and sewing maids, dressmakers. Regardless of the accuracy of such assertions, there is here an implicit acceptance of the ideas of assimilation and acculturation; ideas which have proved central within the modernisation perspective. It is assumed that in the course of service, servants will acquire the value system of their employers and, thereby, the potential to transform their labour force position. Although Broom and Smith's analysis of domestic service is short, their ideas have proved influential in empirical research on contemporary waged domestic labour, particularly within the Latin American context (although see too McBride, 1976).

There are some major drawbacks with the modernisation position which mean that we find this stance seriously deficient, notably the developmentalism inherent in any modernisation argument (Wallerstein, 1979). However, quite apart from the problem of developmentalism, it is the identification of domestic service as the object of analysis which poses severe problems for this perspective in the context of an analysis of waged domestic labour in contemporary Britain. Two problems in particular are insuperable. The first is that for this perspective waged domestic labour has to be about domestic service; modernisation means that this must be so. The second problem is that modernisation is totally incapable of offering an explanation for what would have to be classified as the resurgence of 'domestic service' in the advanced industrialised economies. So, if we attempted to guide our analysis by drawing on this perspective, we could only ultimately move in the direction of a debate with universalist tendencies, in which the key questions would revolve around whether contemporary waged domestic labour represented domestic service revisited. And even this would pose quite major explanatory problems, for it is difficult to see how this 'return' – if found – could begin to be explained by a perspective for which history is but a unilinear progression; or indeed, what descriptions and/or explanations might be offered if the associations with domestic service proved untenable. Grounded firmly in both the developmentalist tradition and a stagist representation of history, and locked into an analysis which would continually hark back to the issue of domestic service, there is little that the modernisation perspective can offer to the analysis of contemporary waged domestic labour in Britain.

Waged domestic labour and segmented labour markets

Accounts of waged domestic labour which utilise the segmented labour market framework are few and far between. However, in a series of articles on Japanese-American women in the San Francisco Bay area of California, Evelyn Glenn develops precisely such a position (1980, 1981,

1986, 1988). The starting point here is the bridging occupation thesis and the ideas of modernisation. However, Glenn's argument is that, contrary to the experience of European women, Japanese-American women in waged domestic labour have experienced occupational ghettoisation, and not a limited form of upward social and economic mobility. Looking at the experience of 'issei' (Japanese picture brides) and 'nisei' (Japanese-American women born in the United States), Glenn shows how waged domestic labour acted as an entry point to the United States labour market for the 'issei', but that 'nisei' women tended to find themselves as concentrated in this occupation as their mothers, if not more so. She goes on to maintain that it is labour market segregation which created this occupational ghetto, effectively trapping successive generations of Japanese-American women until the 1970s.

Similar arguments underpin the analysis of Mary Romero (1988). She suggests that for women of colour waged domestic labour is best conceptualised not as a means towards occupational mobility, but as an occupational ghetto. As a consequence, the research focus shifts to a consideration of how such occupational ghettos come to be reproduced. For Romero, it is only when employers make available to domestics their informal information and contact networks that waged domestic labour can become a bridge towards upward social mobility (ibid.: 89). Her research, however, suggests that this is just what employers do not do: 'domestics do not become "just one of the family". As a result, domestics experience social segregation and do not gain access to other occupations' (ibid.).

There is little doubt that, in comparison to modernisation, the segmented labour market perspective is more promising. It has none of the identification of waged domestic labour with domestic service and its situation within the secondary labour market is acknowledged. Moreover, much of this work recognises explicitly the existence of occupational ghettoisation. But there are general problems with the segmented labour market perspective. The most important of these in this context is its inability to say anything about how and why waged domestic labour has re-emerged in contemporary Britain. Dual and segmented labour market theory, whilst it has some insights to offer into the internal structure of the labour market, is unable to account for broader economic tendencies, for example, the contraction of manufacturing; changes in production systems and the social relations of production; the expansion of the service sector, and so on. The return of waged domestic labour is a trend of this kind. If we followed the segmented labour market route of analysis, serious explanatory difficulties would be encountered. The analysis could only operate at the internal level, that is, at the level of the characteristics of paid domestic workers. And, when we examine the literature on waged domestic labour conducted in this vein, we can see precisely this set of problems. Much is made of the characteristics of paid domestic workers, and of the work, and of occupational ghettoisation, but the explanation for why particular groups of women become and remain paid domestic workers remains stuck at the level of their characteristics as women and as women of colour, with passing reference being made to the lack of alternative labour market openings for these women. All this begs the questions, why are there no alternative labour market opportunities for these women, and why it is that *particular* groups of women, constitute paid domestic workers? It is, in other words, the one-sided emphasis on paid domestic work and worker characteristics which is posing the problems. So, whilst the segmented labour market perspective has definite advantages over modernisation, it is still flawed. Whilst such an approach would stress the characteristics of contemporary waged domestic labour and paid domestic workers, it would be unable either to account for these fully or to unpack the reasons for the contemporary resurgence of waged domestic labour.

Waged domestic labour: a Marxist-feminist approach

As with work within the modernisation perspective, waged domestic labour has for the most part been ignored by Marxist and other broadly left analysts of Third World development, and for similar reasons. Without doubt, this work has been gender blind (Brookfield, 1975; Frank, 1967; Taylor, 1979; Wallerstein, 1979; cf., Edholm, Harris and Young, 1977; Young, 1987), and 'turning the blind eye' has been to do with what count as legitimate objects of study. Indeed, much of the legitimation for ignoring waged domestic labour here has been the non-value productive nature of domestic work; an issue which was well aired in the course of the domestic labour debate (Gardiner, 1975; Seccombe, 1974). By extension, waged domestic labour, as a form of domestic work, can be assumed to have been seen as an unimportant, insignificant economic activity and one worthy of neither detailed comment nor analysis.

The arguments over domestic labour and its relation to the production of value have since been revisited by Marxist-feminists interested in waged domestic labour, and particularly by those working in the South African context (Cock, 1980b, 1981; Gaitskell *et al.*, 1984). In what is the most thorough articulation of waged domestic labour from within a Marxist perspective, Gaitskell *et al.*, argue that waged domestic labour needs to be seen as a special case of domestic work. The key point which they make is that waged domestics, however paltry their wage, are waged workers, and not unwaged housewives. This makes a major difference to conceptualising waged domestic labour. So, whilst the tasks which housewives and domestics perform are identical (cooking, cleaning,

69

washing, childcare etc.), both their position and their status are very different. In short, the social relations of housework are fundamentally different when the domestic worker is waged from those which characterise the unpaid labour of the housewife. When waged, such social relations revolve around control and/or supervision, the ability to hire and fire, and payment for service (Gaitskell *et al.*, 1984: 92). Having said this, the authors accept the general conclusions of the domestic labour debate; domestic labour is considered to be non-productive of value, irrespective of who performs it and how. Given this they argue that neither the housewife nor the domestic can be seen as exploited by capital (cf., Cock, 1980b). However, as befits a Marxist-feminist approach, such conclusions are not taken to signal the end of any interest in waged domestic labour. Instead, Gaitskell *et al.*, identify waged domestic labour as an important site of the triple oppression of class, race and gender. For them, then, one of the problems requiring explanation is how and why particular groups of women end up in waged domestic labour. It is this question which has concerned much of the South African and African literature on waged domestic work.

On a different scale, but also examining waged domestic labour from a Marxist-feminist slant, is Enloe's discussion of the place of waged domestic work within global politics (Enloe, 1989). The focus here is on international labour flows of waged domestic labour, and the argument one which links this with the politics of international debt in general, and the actions of the International Monetary Fund (IMF) in particular. Enloe maintains that the price of the adoption of the IMF package of prescriptive policies weighs more heavily on women than on men. Cutbacks in public expenditure, and specifically in food subsidies, health care and education budgets, are an inevitable feature of IMF support, and mean that the costs of generational and daily social reproduction are carried increasingly by individual households. Given the form of the domestic division of labour, it is women primarily who have to respond to this situation. Ever-increasing levels of resourcefulness are one answer, but for many women paid work becomes essential. The choice which many of these women face is between prostitution, waged domestic work and the Third World production line. For Enloe, then:

> The debt crisis is providing many middle-class women in Britain, Italy, Singapore, Canada, Kuwait and the United States with a new generation of domestic servants. When a woman from Mexico, Jamaica or the Philippines decides to emigrate in order to make money as a domestic servant, she is designing her own international debt politics. She is trying to cope with the loss of earning power and the rise in the cost of living at home by cleaning bathrooms in the countries of the bankers.
>
> (Enloe, 1989: 185)

In comparison with the modernisation and segmented labour market perspectives, the Marxist-feminist perspective provides a sound analysis of waged domestic labour as a general economic phenomenon. Dependent upon cash payments made out of the disposable income of households, waged domestic labour, like all forms of domestic labour, is shown to be a non-productive form of labour, the outcome of which is the daily and generational reproduction of households and labour power. Beyond this, the Marxist-feminist perspective also begins to develop some vital connections between paid domestic work, domestic labour and the household. The household is conceptualised within this perspective as a site of social reproduction, rather than simply as the locus for paid domestic work. Domestic labour is seen as central to this, but it is acknowledged that it can take a number of forms. The unwaged form is perhaps the most common version, but waged domestic labour represents another. And both of these forms are seen as characterised by different social relations. The importance of such connections is not to be underestimated. In analytical terms it is imperative that waged domestic labour be conceptualised as a particular form of domestic labour. Indeed, without such a conceptualisation it is impossible to move beyond accounts which remain stuck at the level of paid domestic workers and their characteristics or which see waged domestic labour as historically unique. Moreover, without a connection with the general category, 'domestic labour', it is impossible to relate waged domestic labour to the unwaged form of domestic labour. This relation is vital because one of the fundamental tasks of any theoretical work in this area must be to account for how and why in particular periods and places domestic labour is arranged as it is. In Britain for much of the post-war period, domestic labour has been unwaged and performed by the housewife. Nowadays the employment of certain categories of domestic labour is becoming increasingly commonplace. In order to account for such a transformation, then, we need a perspective which is sensitive to the idea of the general category 'domestic labour' and its particular forms. The Marxist-feminist perspective certainly is this. Furthermore, using the notion of individual instances, it is able to cope with variation in the unwaged and waged forms of domestic labour. Seen through this lens, nineteenth-century domestic service in its populist 'Upstairs–Downstairs' conceptualisation represents one instance of the waged form of domestic labour, with its own unique master–servant social relations. Twentieth-century paid domestic work in South Africa, with a very different form of social relations (grounded in racism and the colonial legacy), represents another. Waged domestic labour in the contemporary United States, where immigration and the sponsor relation figure so highly in the social relations of work, can be seen to constitute yet a further instance of waged domestic labour, and so on and so forth.

71

In all this, however, there are also problems. Whilst the Marxist-feminist perspective is perfectly capable of formulating waged domestic labour as a particular form of domestic labour, and of accommodating the notion of individual instances, it runs into two key difficulties. The first is in attempting to explain why the contemporary transformation from unwaged to waged domestic labour has come about. The second is a standard line of criticism, that such a perspective has nothing to contribute to an explanation of the heavily gendered nature of waged domestic labour. Of the two, the former is perhaps the most fundamental problem. Whilst it was possible to explain the historical decline of domestic service in terms of the expansion of the productive forces and the proletarianisation of women, the resurgence of waged domestic labour in the advanced industrial countries in particular is more difficult to cope with. The Marxist-feminist position, then, offers some useful insights into thinking about domestic labour generally. It also enables us to focus on particular forms of waged domestic labour, each with particular sets of social relations. We see this representation as a useful and necessary one. However, in terms of actual development, it is the remaining theoretical perspectives which have the most to offer.

Waged domestic labour and gender ideology

A fourth conceptualisation of waged domestic labour is one developed by feminist historians, most notably by Leonore Davidoff (1973a and b, 1974, 1983), Patricia Branca (1975) and Martha Vicinus (1972). Davidoff's starting point is to argue that social divisions are most clearly visible in the reproductive sphere, and that it is to the household therefore that historians should direct their attentions if they are fully to understand Victorian Britain. What emerges from this focus is a clear picture of Victorian gender ideology; of the gender identities of Victorian middle-class women; and of the role of domestic service in enabling the reproduction of these identities. There is little doubt that, throughout Victorian Britain, the dominant gender ideology was one which saw the home as 'the natural place for women' (Davidoff, 1974: 417). However, particular versions of this ideology can be seen to have been shaped both by class and by the institution of domestic service. For middle-class women, the home was the hearth and a focus of conspicuous consumption (Vicinus, 1972: ix). Middle-class women were 'the keepers of the hearth', household decision-makers and managers (Branca, 1975: 18), as well as guardians of normative social mores. In comparison, for working-class women, the middle-class home represented a place of hard physical labour. These differences, Davidoff argues, were manifested in the social and spatial divisions visible within the middle-class household, and specifically in the segregation between servant space and household space. They were also reinforced by servants' protection of the middle classes from 'defiling contact with the sordid or disordered parts of life' (Davidoff, 1974: 412). For example, servants mediated the middle classes' contact with the world both beyond the household and within – a pattern of social organisation which enabled the reproduction of 'the home as haven' ideology. In a different way – in that domestic service, it was argued, would provide training for later married life – domestic service is shown to have assisted in the reproduction of working-class women's identification with the home as a place of physical work. Using the diaries of Hannah Cullwick, Davidoff has gone on to show how domestic service played a key role in the construction of Victorian views on sexuality (Davidoff, 1983). The institution of domestic service and its social relations are seen to have enabled the separation of the middle classes from physicality, the body and bodily functions, as well as the establishment of a dual representation of women, in which working-class women, by virtue of their associations with the body (through domestic work) were seen to be closer to nature than their middle-class sisters. In turn, Davidoff maintains this brought about a closer identification between working-class women and sexuality than was the case for middle-class women – a theme which has been expressed traditionally in terms of the sexual relations between servants and male household members (Gillis, 1983).

Waged domestic labour and role strain theory

Within the literature on dual-career couples, as we have seen, waged domestic labour is conceptualised predominantly in terms of role strain theory. Stress and strain are seen as an inevitable part of the dual-career household; their management as necessary for the continued reproduction of this 'new' family form. Domestic work, this literature acknowledges, poses a potential 'overload dilemma' for dual-career couples (Rapoport and Rapoport, 1969). Moreover, others, for instance Holmstrom (1972) and Hunt and Hunt (1982), have pointed to the socio-psychological stress which a failure to keep on top of domestic work can induce, particularly for the female partner in the dual-career couple. Waged domestic labour is conceptualised here as a coping strategy (or 'couple strategy'), that is, as a mechanism to alleviate physical overload and psychological stress.

Waged domestic labour and the changing occupational class structure

A final conceptualisation of waged domestic labour, and one also developed within the literature on the dual-career couple, is that sketched by Rosanna Hertz (1986). Although chiefly concerned with investigating the dual-career couple, Hertz clearly sees waged domestic labour as one of the key contingencies in the reproduction of this type of household (see above, p. 64). However, she also sees the availability of waged domestic labour – as well as the existence of the dual-career couple itself – as the product of the occupational class structure. Indeed, although she does not elaborate further, Hertz maintains that the current occupational class structure produces two groups in relation to domestic work and childcare; those who have not got the time to do it, and those who have no alternative but to do it.

Although thinly sketched, Hertz's ideas on the importance of the occupational class structure have considerable potential. Changes in the occupational class structure in Britain, in particular the growth of professional and managerial employment, and their mediation by gender, together with the increasing prevalence of the dual-career household, have created the conditions for a potential crisis in daily social reproduction. At the same time, other changes in the occupational class structure, notably the decline in manual work and the commodification of previously unwaged forms of 'domestic' work (Watson, 1991), have created new forms of waged work 'for women'. We expand on this position in the following two chapters. However, such conditions are only potential ones. To account fully for the increasing prevalence of waged domestic labour in contemporary Britain, we also have to explain why individual households switch to using waged domestic workers and why other people work in this capacity. In the first case, whilst the role strain theory associated with much of the literature on the dual-career family has many of the difficulties associated with role theory generally (see for instance, Connell, 1987), the notion of coping strategies is still useful. As we will argue in the following chapter, this offers an initial inroad into examining how and why individual households move to using waged domestic labour. None the less, whilst coping strategies work from the notions of stress and stress management, they neither provide a handle on why (and how) some households come to see domestic work as a source of coping difficulties, nor do they explain how and why waged domestic labour comes to be seen as the solution to this problem. It is at this level that the work of feminist historians on gender ideology and nineteenth-century domestic service has something important to contribute.

The writing of Davidoff, in particular, has shown how the particular form of domestic labour in Victorian Britain played a key role in the construction of both middle-class and working-class women's identities as women. Domestic labour has not stopped being important in this process of identity formation. For most of the post-war period, the concept and reality of being a housewife has occupied a pivotal role for the majority of women in Britain (Oakley, 1974). Both for society, and for women, domestic labour in its unwaged form has symbolised mothering, nurturing and caring, has constituted women's traditional gender role, and has provided one of the fundamental sites through which patriarchal gender relations are constructed. Such ideas and realities underpin why domestic labour, and coping with its demands, continue to be seen as important by women, and why, notwithstanding twenty years of second wave feminism, domestic labour remains overwhelmingly women's work. It is a societal expectation that domestic labour simply has to be done, a normative expectation and one which carries with it heavy gender connotations. Coping crises with respect to domestic labour therefore cannot be ignored. But, to resolve things using waged domestic labour requires that certain women (and men) in Britain have challenged at least some of the assumptions and associations which have served to underpin the hegemonic post-war form of domestic labour as unwaged work.

In the following chapter we explore the conditions which have given rise to the contemporary resurgence of waged domestic labour in Britain, as well as how and why individual households have come to see the employment of waged domestic labour as a necessary aspect of social reproduction.

4

UNPACKING DEMAND

In this chapter we account for the resurgence in demand for waged domestic labour generally in Britain through the 1980s amongst the new middle classes. The focus – for reasons relating to the profusion of categories of waged domestic labour revealed in Chapter Two – is not on waged domestic labour as a general category, but on two particular categories of waged domestic labour. The demand encountered included over 100 categories of waged domestic labour. In such circumstances it is theoretically inappropriate to think of waged domestic labour as an homogeneous occupational category, as well as inaccurate to account for trends in its utilisation in a general sense. Yet, as we have seen, the resurgence in demand for waged domestic labour in Britain in the 1980s and the major categories of demand through the same period were primarily bound up with the middle classes. Given this, an analysis of the resurgence in waged domestic labour in Britain through the 1980s must focus on those categories for which demand was particularly buoyant as well as the source of this demand. In this chapter, therefore, we concentrate on the two forms of waged domestic labour most closely associated with the new middle classes throughout the 1980s, namely the cleaner and the nanny, and for which demand was particularly high, specifically in the late 1980s.

We commence this chapter, however, with a number of comments which relate to waged domestic labour as a general occupational category. This enables us to situate waged domestic labour within the service sector industries and to establish the explanatory form of our argument. As will become clear, we consider the resurgence in demand for waged domestic labour generally, and the expansion in demand for cleaners and nannies in contemporary Britain, to be highly contingent phenomena. Our explanation therefore is one which focuses on establishing the historical conjuncture of trends and circumstances in the 1980s which provided the conditions in which middle-class households could see the need to substitute these particular categories of waged domestic labour for their own unwaged household labour.

WAGED DOMESTIC LABOUR AND THE SERVICE SECTOR

As something involving the servicing of individuals and households, waged domestic labour is an activity which has to be categorised as a personal service. Moreover, it is, for the most part, a service activity which is purchased directly by households. As we showed in Chapter Two, whilst the 1980s saw a considerable expansion in both the number of firms and employment agencies specialising in private domestic services, the majority of households (and most of our respondent households) purchase(d) their waged domestic labour directly, either through formal or informal advertising. Given this, waged domestic labour needs to be seen predominantly as a consumer service activity, rather than as an activity located within either the sphere of production or the sphere of circulation (Allen, 1988a).

Whilst much attention has been given over recently to the service sector and to its growth, importance and conceptualisation (Allen, 1988b; Buck, 1985; Damesick, 1986; Gershuny and Miles, 1983; Rajan, 1987; Urry, 1987), most of this attention has been focused on producer services (see, for instance, Allen, 1988a; Daniels, 1986; Daniels, Leyshon and Thrift, 1986; Gillespie and Green, 1987; N. Marshall, 1985, Marshall, Damesick and Wood, 1987; Wood, 1986). In comparison, little attention has been accorded to consumer services. Indeed, within this latter group it is the predominantly public sector services of health and education, rather than the 'other' personal service industries – for example, hairdressers, dry cleaners, shoe repairers, etc. – which have received most consideration.

It is not difficult to see why the literature on services has tended to ignore the 'other' personal services category. In terms of employment numbers (the problematical, but widely used, index of service sector output), 1981 figures showed a marked contraction (a 41.3 per cent decline over the period 1959–81; see Allen, 1988b: 99). In contrast, other service industries – notably welfare, health and education, and financial and business services – exhibited spectacular increases (see Allen, 1988b: 98). Given this, we could explain the lack of concern for the 'other' personal service industries simply in terms of employment trends. However, whilst this certainly has been influential, the 'other' personal service industries are regarded as insignificant by the major theoretical perspectives on services. For example, post-industrial society arguments emphasise three phases of economic development; the pre-industrial, the industrial and the post-industrial (D. Bell, 1973). It is, none the less, only within the pre-industrial phase that the 'other' personal services category is seen to be one of the major employment categories – largely because of the importance of domestic service. In contrast, the arguments made by Gershuny (1978; and see too Gershuny and Miles, 1983) regarding the development of the 'self-service' economy stress the labour intensive characteristics of many of the personal service industries. According to Gershuny, this has led to a situation in which households have increasingly substituted manufactured goods for personal services – washing machines for laundries; cars for public transport; videos for cinema visits, and so on. Viewed through this lens, the contraction of the 'other' personal service industries is seen as a long-term trend, and is explained in terms of the growth of a 'self-service' economy based on technological innovation and the widespread ownership of material commodities. Finally, the Marxist perspective on services also pays minimal attention to the 'other' personal service industries (Mandel, 1975). The expansion in service sector employment is here interpreted in terms of a move by capital into new, more profitable investments and activities, some of them within the spheres of reproduction and circulation (for example, retail services, hotels and catering, and leisure services). At the same time, developments within manufacturing are seen as enabling households to replace labour intensive personal services with manufactured goods. Once more, the 'other' personal service industries are seen as in decline; although it is acknowledged that the trend of 'contracting out' in manufacturing and certain of the public sector services has led to an expansion in activities such as industrial contract cleaning. But, of greater interest are the industries and activities into which capital is expanding – for example, tourism and the leisure and cultural industries (Urry, 1990) – as well as the proliferation of employment in business and financial services.

Clearly the reasons why the 'other' personal service industries have been overlooked in contemporary debate on the service sector are powerful ones. Not only have these industries frequently experienced a major decline in employment terms, but all the theoretical perspectives on services argue that this decline is long-term and fundamental. This poses some major explanatory problems for the situation described in Chapter Two: how are we to account for the resurgence of an occupational category which is part of an apparently declining group of services? The explanation which we put forward in this chapter is fundamentally a contingent one, and one which owes much to the complexities revealed by our empirical investigations within middle-class households. However, we would argue further: that, given the general underlying trend of employment decline in the 'other' personal service industries, any explanation for the resurgence of waged domestic labour must be contingent. If employment in most of these 'other' personal service industries is in decline, and employment in private domestic services is not, then there must be something particular to domestic work itself – and not to the 'other' personal services – which is behind this expansion. We are convinced that the critical issue here is that there are certain aspects of domestic work, certain specific domestic tasks, which are not amenable to self-provisioning through the purchase and use of manufactured goods. Instead, these tasks remain labour intensive. Combine this with the gendering of specific household tasks and certain other key contingencies, notably the mediation of the occupational class structure by gender, the form of contemporary childcare provision and the growth of leisure activities within the middle classes, and the potential for the expansion in employment in private domestic services is readily understandable. The first part of this chapter considers these particular contingencies in some depth, as well as the mode of domestic provisioning through the post-war period and the way in which this is mediated by class, gender and domestic tasks. It is this conjuncture in Britain in the 1980s which we consider created the conditions in which middle-class households could see a need to substitute certain categories of waged domestic labour, specifically cleaners and nannies, for their own unwaged domestic labour.

THE MODE OF DOMESTIC SERVICE PROVISION AND ITS MEDIATION BY GENDER AND CLASS

Our starting point in this analysis is the nature of domestic work itself. We begin by examining the way in which households' needs for domestic services have been met within the post-war period. In this respect the most significant arguments are those of Gershuny (1978, 1985) and Gershuny and Miles (1983); (although see too the arguments of Pahl (1984) on the growth of 'self-provisioning' activities within households).

Domestic needs are basic ones. Domestic labour, whether this be cleaning, washing or cooking, is fundamental to individual and social reproduction, both in a day-to-day sense and generationally. Traditionally and historically in Britain, such needs have been met through the full-time labour of women, either in an unwaged capacity as housewives (Lewis, 1980; Oakley, 1974) or as waged domestic servants (see Chapter Three). A clearly defined gender division of labour therefore is widely acknowledged to have underpinned the historical and contemporary organisation of domestic work. In the post-war period, as Gershuny's arguments have illustrated, technological innovation has played a key role, both in changing the characteristics of many domestic tasks, and in shaping the amount of labour time devoted to them. The advent of 'labour-saving' appliances and commodities (for example, washing machines, convenience foods and timer ovens and – more recently – washer-drier machines, slow-cookers, microwaves and dishwashers) together with their relative low cost in a period of generally rising household incomes (Pond, 1989), is widely heralded as bringing about a qualitative and quantitative change in the nature of housework in the post-war period in Britain. At the heart of this change is a reduction in the amount of labour time spent on domestic work; a trend which is extensively, although varyingly, invoked in many explanations of the growth of women's employment through the same period. However, as Gershuny shows, whilst this trend applies to working-class households, it does not characterise what was happening in middle-class households in the early years of this era. Indeed, the years 1937–61 saw an increase in the amount of middle-class female labour time devoted to domestic work (Gershuny, 1985: 150–1). For Gershuny, the reason for these class differences is the still widespread use of servants by the middle classes in the inter-war years and their subsequent decline in the post-war period.

Gershuny's arguments regarding the mode of domestic service provision in Britain in the post-war period, although important and often convincing, are none the less flawed, at least in part. For our purposes the most significant oversight occurs in Gershuny's assumption that technology mediates *all* forms of domestic labour. Such assumptions can be seen to underpin the following statement, 'The same, or at least analogous, needs that are satisfied by domestic servants, train tickets and theatre seats are met later by domestic machinery, motor cars and video recorders' (Gershuny, 1985: 129). The first problem with this assumption is that not all forms of domestic labour are capable of being performed exclusively by domestic machinery (and see too Cowan, 1983). The most obvious area of difficulty in this respect is household cleaning. In spite of widespread household ownership of vacuum cleaners and purchasing of commodities such as dusters, cleaning cloths, mops and cleaning products, keeping the house clean still takes a considerable amount of physical labour time. Ironing exhibits similar characteristics. Notwithstanding developments in fabrics, clothing and washing machine technology, ironing – like cleaning – remains a labour intensive task. And then there is childcare, or rather what we prefer to see as childcare activities (Gregson and Lowe, 1993). Again, these are inherently, and necessarily, labour intensive. Certain basic household tasks then remain labour intensive. Moreover, given the nature of these tasks, it is difficult to envisage them as anything else.

The second problem with Gershuny's assumption that technology mediates all forms of domestic labour is its inherent developmentalism. Technological innovation with respect to domestic service provision is assumed to be progressive; diffusion through households, inevitable and irreversible. What this perspective ignores is that domestic labour is socially, and not technologically, constructed. As feminist work on housework has long recognised, domestic tasks, the products of domestic labour, the commodities used in their production, as well as how to do domestic tasks and how to make domestic products, are all shaped by the ideas and values present within specific societies. For the most part, such arguments have focused on the way in which gender mediates the above, and particularly on how the above come to be gendered activities. But, within the context of contemporary Britain, it is possible to detect a class-based move away from various aspects of the technological fix with respect to housework. A range of environmentally based social movements (for example, vegetarianism and a host of broadly 'green' movements) has generated a more critical stance towards convenience foods and microwave ovens. Wholefoods, 'bake your own' and traditional cooking methods are currently in vogue in the food preparation sphere. In clothing – largely in response to trends within the textile and clothing industries – there has been a reaction against artificial fibres and the return to favour of the 'natural' fibres (cotton, linen, silk, wool). Such fabrics require careful washing, (often by hand) and, in some cases, steam ironing. And in cleaning, the late 1980s and early 1990s saw the introduction of the new range of 'eco' products. Purchased it seems almost exclusively by the middle classes, such products are generally seen to involve twice as much in the way of 'elbow grease' as their less environmentally friendly competitors. Add to this various trends in household interior decoration (for example, traditional pine kitchens, stripped doors and exposed wooden floorboards) and the message is clear. Through the 1980s in Britain, a variety of trends conjoined to increase potentially the amount of labour time which individual household members could devote to certain essential domestic tasks. Rather than domestic labour being carried out increasingly and inevitably by commodities produced within the sphere of production, such tasks – and the form which they take in particular households – are mediated by class and by social movements, by gender and by age, as well as by technological innovations within the household and consumer goods sectors.

Clearly our arguments concerning relative amounts of time spent on domestic labour are speculative. Indeed, they would require longitudinal and historical time budget data to confirm fully – a project way beyond the capacity of this research. Nevertheless, market research findings provide support for some of the arguments made above, particularly those regarding certain of the labour intensive tasks, notably cleaning. Thus, for example, a MORI survey conducted in 1990 suggested that 37 per cent of all women in Britain spend over ten hours per week cleaning, with another 28 per cent spending between six and ten hours on cleaning tasks.[1] In addition, the same survey revealed that 77 per cent of all women vacuum every two to three days; that 75 per cent dust every two to three days and that 66 per cent clean the toilet every two to three days. Such findings provide ample testimony for our argument that, notwithstanding the widespread ownership of a variety of forms of domestic cleaning appliances and commodities, the amount of labour time devoted to household cleaning tasks in contemporary Britain is still considerable.[2] When we combine these findings with what we know about the gender characteristics of these labour intensive household tasks, the importance of these arguments concerning labour time becomes more apparent.

Whilst it has long been recognised that, in an aggregate sense, women in Britain perform the bulk of domestic labour (Oakley, 1974), it is only recently that the gendering of *particular* domestic tasks has been acknowledged and investigated (Pahl, 1984; Wheelock, 1990). One of the most important points to emerge from this research on domestic work concerns the finding that domestic labour comprises both gender-segregated and gender-neutral tasks. Thus, in Wheelock's research on north-east households with unemployed male partners and female partners in part-time employment, washing, ironing and 'proper' cleaning were identified as tasks which were clearly the responsibility of women (Wheelock, 1990).[3] In contrast, other tasks – for example, vacuuming, cooking and childcare – were constructed as shared activities. A further important feature of Wheelock's research is her differentiation between domestic tasks on the basis of the labour time which each requires. In making this point, Wheelock is recognising the relative importance of individual domestic tasks to daily and generational reproduction. This we feel to be critical to the advancement of debate on the form of the domestic division of labour in contemporary Britain (Gregson and Lowe, 1993, 1994). The categories which Wheelock works with are major regular, major irregular, minor regular and minor irregular and the point which emerges when categorisations on the basis of labour time are combined with those of gender is that a significant number of the major regular domestic tasks are also those which are the most strongly gender-segregated. In particular, cleaning, washing and ironing are shown to be both labour intensive and, almost invariably, 'women-only' tasks.

Wheelock's research, though based on a rather unconventional household form, is, none the less, concerned with 'working-class' households. As part of our investigations on middle-class households, we examined the extent to which Wheelock's findings were replicated within households with partners in very different class positions. Our results proved remarkably similar (Gregson and Lowe, 1993). Household washing was performed exclusively by women in 81 per cent of our north-east households and 71 per cent of our south-east households. Ironing was segregated to a similar degree, with 70 per cent of both our north-east and south-east households revealing the female partner to be solely responsible. Dusting, bathroom cleaning and floor cleaning (excluding vacuuming) – like washing and ironing, major regular tasks – were also strongly gendered as female tasks.[4] In contrast, vacuuming, food shopping, cooking and washing up were identifiably shared tasks, as were some of the more traditionally 'male' tasks, such as decorating and gardening. In general then our findings suggest that the pattern of gender-segregated and gender-neutral domestic tasks is common across both social class and space in contemporary Britain. Moreover, it would appear that within middle-class households, whilst the traditionally 'male', as well as some of the major regular domestic tasks, are becoming increasingly shared, a significant number of the major regular tasks remain exclusively 'female'. When we link this second point back to our critique of Gershuny, it is clear that these segregated tasks (washing, ironing and cleaning) are the very tasks which are labour intensive, are likely to remain so and may, indeed, also be currently more demanding than previously in terms of labour time.

We consider the above situation to be of particular importance in accounting for the high levels of demand for the cleaning category of waged domestic labour by middle-class households in Britain through the 1980s. Our findings show that middle-class women in Britain – even when in full-time service class occupations – bear full responsibility for the most labour intensive of domestic tasks (i.e., cleaning and ironing), as well as an increasing share of responsibility for the traditionally 'male' tasks and primary responsibility for the remaining tasks (for a full discussion of this see Gregson and Lowe, 1993). Furthermore, if – as we suspect – the societal construction of domestic tasks has altered in the course of the 1980s to mean that certain tasks, at least, are potentially more demanding in the way of labour time, then this means that the labour time which individual middle-class women need to spend on certain labour intensive domestic tasks has either potentially increased, or remains considerable.[5] It is this situation which provides the most appropriate backdrop against which to consider the expansion in women's employment within the 'service class' occupations through the 1980s.

THE EMERGING 'SERVICE CLASS' AND ITS MEDIATION BY GENDER

The nature of the occupational class structure and its mediation by gender is a vast, complex field, with a literature far too extensive to be surveyed fully here (see, for instance, Britten and Heath, 1983; Crompton, 1989a; Crompton and Mann, 1986; Gamarnikow *et al.*, 1983; Goldthorpe, 1983; Lockwood, 1986; Marshall *et al.*, 1988; Wright, 1985). However, one trend which has been identified and debated at length within the literature is the emergence of the 'service class' or 'new middle-class'. It is none the less only recently that attention has been directed to the gendering and gender characteristics of the occupations which comprise the so-called service class, that is, professional, managerial and high grade administrative work.

Following Witz (1992: 204), it can be stated that the concept of the service class is something which gained increasingly in popularity through the 1980s. Indeed, the term is used frequently in most analyses of the contemporary occupational class structure in Britain (Abercrombie and Urry, 1983; Goldthorpe, 1982; Lash and Urry, 1987; Thrift, 1987, 1989; cf. Savage et al., 1992). Although originating in the traditions of Austrian Marxism (where it was first used in 1924 by Karl Renner to refer to the rise of the professional, as opposed to owner, manager), the term reappeared in Dahrendorf's work on class and class conflict, where it was used to describe 'salaried employees who occupy positions that are part of a bureaucratic hierarchy' (Dahrendorf, 1959: 55). More recently, it has resurfaced in both sociology and human geography, and is used to refer to 'higher level white collar occupational groups in advantaged positions in modern societies: namely professional, administrative and managerial employees' (Witz, 1992: 205).

There is much that is problematical about the concept of the service class, particularly once the term is used in anything other than a descriptive sense. It is, for example, difficult to theorise as a coherent collective entity – is it *the* service class, or something much more fragmentary (Savage et al., 1988; Savage et al., 1992)? Moreover, how does it relate to the middle classes? And if it is, as we feel, so fragmentary, combining individuals with vastly different sets of assets and skills, what does this mean for questions of status, identity and consciousness? At the definitional level there is the boundary problem. Just where does one draw the line between higher and lower service class occupations? And, as Witz has argued recently, there is in all this an emphasis on the gender characteristics of the service class, rather than the place of gender in the construction and constitution of contemporary service class positions (Witz, 1992: 205). All these are fascinating and important questions, but for our purposes the key immediate issues concern the degree and nature of women's participation within high-level professional, administrative and managerial occupations, and the way in which this relates to domestic labour. We therefore use the term service class in a purely descriptive sense, as shorthand for employment in a particular, heterogeneous cluster of occupations.

The significance of service class occupations within the contemporary British occupational class structure, and their expansion relative to manual work, is well documented (Crompton and Sanderson, 1990; Savage et al., 1988; Thrift, 1987). Table 4.1 records the number and

Table 4.1 Gender composition of service class employment, 1971–81

SEG no.	Description	Men	Women	Total	1971	1981
		(1) Percentage change 1971–81			*(2)* Percentage distribution of total (men and women) by column	
1	Employers, managers, etc., in industry and commerce: large establishments;					
2	Employers, managers, etc., in industry and commerce: small establishments;	+16	+45	+20	12.5	14.8
3	Professional workers, self-employed;					
4	Professional workers, employees					
5	Ancillary workers, artists; foremen & supervisors – non-manual	+28	+41	+35	7.6	10.1
6	Junior non-manual workers	−32	8	−3	21.2	20.3
7	Personal service workers	+12	+5	+6	5.2	5.4
8	Foremen and supervisors – manual;					
9	Skilled manual workers;	−11	−11	−11	44.0	38.6
10	Semi-skilled manual workers;					
11	Unskilled manual workers					
12	Own account workers (other than professional)	+17	+3	+14	3.5	4.0
13	Farmers – employers and managers;					
14	Farmers–own account	−19	+18	+19	2.4	1.9
15	Agricultural workers					
16	Members of armed forces	+1	+50	+4	1.0	1.0
17	Inadequately described and not stated occupations	–	–	–	2.6	3.8
	nos				100 2,503,143	100 2,540,559

Source: Crompton, R. and Sanderson, K. (1990) *Gendered Jobs and Social Change*, London: Unwin Hyman, p. 164

percentage of women and men working in high-level service class occupations in 1971 and 1981. Between 1971 and 1981 the percentage of women working in professional and managerial occupations (i.e., SEGs 1–4 inclusive) increased from 16 per cent of all SEG 1–4 employees to 19 per cent, a 45 per cent increase (Crompton and Sanderson,

1990: 165), and it is anticipated that this trend continued through the 1980s into the 1990s. Further confirmation of the changing gender characteristics of professional and managerial occupations is given by occupation-specific data:

> Women were 30 per cent of first enrolments in medicine in 1970 but 45 per cent by 1983; 7 per cent of new members of the Institute of Chartered Accountants in 1975, but 23 per cent by 1984; 2 per cent of successful candidates in the final examinations of the Institute of Bankers in 1970 but 21 per cent by 1983; and 19 per cent of successful Law Society finalists in 1975, but 47 per cent by 1984.
>
> (Crompton and Sanderson, 1990: 69)

Whilst absolute and relative figures on women's participation within the upper echelons of the service class are significant, it is the nature of their employment within these occupations which is of paramount importance to our interests in waged domestic labour. Some of the most interesting and formative work in this respect is Crompton and Sanderson's research on pharmacy and accountancy (Crompton and Sanderson, 1990; and see too Savage and Witz, 1992). One of the central arguments put forward by Crompton and Sanderson is that those occupations which use rational and legalistic strategies (i.e., the 'qualifications lever') to control access to their ranks (typically the professions) are those which have proved most vulnerable to the gradual and continuous expansion in women's educational attainment (Crompton and Sanderson, 1990: 64). However, Witz's work on the emerging division of medical labour in the late nineteenth and early twentieth centuries adds some important qualifications to this. Although Crompton and Sanderson acknowledge that increasing levels of educational attainment and/or credentials gained by women will not automatically be reflected in gender equality within career-structured occupations, they still maintain the possibility of non-gendered patterns of career-specific advancement (Crompton and Sanderson, 1990: 166). Witz's research, though, shows that access is one thing and the paths of male and female advancement quite another. Here strategies of exclusion (principally the qualifications barrier) are revealed to be only one amongst a number of gendered projects of professional closure (Witz, 1992).

At one level we can read Crompton and Sanderson's work as indicating the importance of particular occupations, and their specific career structures, in mediating the shape of women's participation within professional and managerial employment (and see too Halford, Savage and Witz, forthcoming). In the early post-war period, particularly during the 1950s and 1960s, women's professional employment was limited mainly to teaching and to nursing. This situation reflected the qualifications barrier, particularly women's lack of access to higher education in the university sector. Women's entry into pharmacy however has mirrored exactly the teaching/nursing pattern. As in teaching and nursing, women are concentrated at the practitioner level and are frequently employed part-time (see also Podmore and Spencer (1986) and Abel (1989) on the patterns of female participation within the legal profession). Such findings have been used by Savage *et al.* (1992) to suggest that the expansion of the professional division of labour, particularly through the 1980s, and its associated processes of professional demarcation, is strongly gendered, with women tending to be employed in subordinate positions (Savage *et al.*, 1992: 76–7; Savage *et al.*, 1992). We would agree in general with these arguments, but we feel that it is important not to lose sight of the fact that women's entry into the professions through the 1980s has increased, and that the range of professions within which they are employed has expanded. We take the latter to be particularly significant. Indeed, there are now various professions to which women have gained entry (notably the newer, organisationally structured ones, such as accountancy) and within which their pattern of participation (if not the level which they eventually attain) is identical to that of men, i.e., full-time. It is the pattern of female participation within at least certain of these professional/managerial occupations which poses major problems for the traditional form of the domestic division of labour within the household and it is here that the developments within women's employment outlined above intersect with the arguments of the previous section.

Crompton and Sanderson themselves point, in passing, to the incompatibility of full-time service class employment with the conventional female domestic role:

> As women begin to qualify and enter into other occupations the complementarity between paid work and domestic responsibilities, such as is found in teaching, can no longer be assumed. The working day in a bank or insurance company is not structured around a timetable appropriate to the demands of a husband and family. Indeed, it may be suggested that the difficulties faced by such women . . . may be a source of pressure for change in the division of labour in both the narrower and broader senses.
>
> (Crompton and Sanderson, 1990: 58)

The immediate point which Crompton and Sanderson are making here, of course, is to do with the sheer physical impossibility of one person combining full-time professional/managerial employment with the conventional full-time female domestic role of wife/mother/domestic labourer, at least on anything other than a short-term basis. Real though such physical difficulties undoubtedly are, however, it is the differential

87

structuring of the day by, on the one hand, service class occupations and, on the other, the institutions which shape the activities of children and service households' needs (for example, schools, doctors, dentists and shops) which are taken to exacerbate the problem. For those households where both partners' pattern of participation in paid work is shaped by the time–space demands of service class occupations, major problems of time–space synchronisation are anticipated. In such circumstances, some form of change in the domestic division of labour is widely held to be inevitable.

Whilst we agree wholeheartedly with Crompton and Sanderson that the conflict between different time–space structures produces major problems for partners in service class occupations, we are less than convinced that these are as productive in producing pressure for change in the domestic division of labour as Crompton and Sanderson seem to believe. Instead, we maintain that this argument is fundamentally flawed. As with most research on paid work, Crompton and Sanderson operate with a black box view of the household (Morris, 1991) and an equally monolithic approach to domestic labour. They see changes in women's participation in paid employment as having a potential but straightforward effect on the organisation of domestic work. Such arguments have been shown to be seriously deficient by the most recent research on the organisation of domestic work within the household. Morris's research on working-class households in South Wales, for example, has shown that no necessary connection exists between what happens to men and women in the worlds of (un)employment and the form of the domestic division of labour (Morris, 1985a and b). Likewise, our own investigations into the form of the domestic division of labour within middle-class households with both partners in full-time service class occupations corroborates certain, if not all, of Morris's arguments and, in particular, the resilience of the traditional form of the domestic division of labour (Gregson and Lowe, 1993). Notwithstanding the force of these arguments, though, it is the equally black-box approach to domestic labour which is of greater significance.

Returning to the arguments made in the previous section, our research has suggested that some important developments are occurring in the form of the domestic division of labour within a significant proportion of service class households; notably a tendency for partners to share the traditional 'male' tasks, a number of the minor regular tasks (including food shopping and the school run) as well as some of the major regular 'female' tasks (particularly cooking and vacuuming). But, as we demonstrated above, the most labour intensive of domestic tasks remain exclusively 'female'. The sharing between partners of certain household tasks indicates that the conflicting time–space demands of institutions, in some cases at least, are being met by both partners, and not by one. Nevertheless, awkward though they are, such tasks are only problematical in that they require the resolution of conflicting time–space structures. They do not consume vast amounts of labour time. In contrast, the more labour intensive tasks are problematical for partners precisely because they require them to find actual labour time in quantity – as market research has shown, as much as ten hours per week for household cleaning and a period of hours, too, for household ironing. *It follows therefore that it is not domestic labour in general which is the problem for households where both partners are in full-time professional, managerial or administrative occupations. Rather it is the labour intensive tasks of cleaning and ironing in particular which are the major problem. And, given the gendered nature of these tasks, it is women in service class occupations who, for the most part, have to try to synchronise the time–space demands of paid employment with those of household needs for these particular domestic services.* When, therefore, we combine our arguments about the mode of domestic service provision and its mediation by class and gender with the expansion of, and potential further growth in, women's employment in high-level service class occupations, and particularly with the varying patterns of female participation within these, then it is clear that for some middle-class women in the 1980s in Britain certain of the labour intensive major regular domestic tasks posed considerable problems. We see this conjuncture of trends and circumstances as particularly influential in bringing about at least some of the expansion in demand for the cleaner category of waged domestic labour amongst middle-class households through this period.

Thus far, we have side-stepped the issue of childcare. Childcare, however, and particularly the need of service class households for it, is one of the major forces behind the rise of waged domestic labour in contemporary Britain. In the following section we focus specifically on this.

CHILDCARE PROVISION IN POST-WAR BRITAIN

The nanny evolved among the upper and upper-middle classes during the nineteenth century, flourished for approximately eighty years and then, with the Second World War, disappeared forever.
(Gathorne-Hardy, 1972: 7)

Although estimates of the number of nannies currently employed in households in Britain are notoriously difficult to make, our work (see Chapter Two), that of Cohen (1988) and Femiola (1992), and the 1991 General Household Survey show the nanny to be an important feature of contemporary childcare provision in Britain, albeit for a minority of households, most of them in service class occupations (Mottershead, 1988).[6] Gathorne-Hardy's pronouncements therefore have proved to be short sighted. Why? In part the answer is suggested by Gathorne-Hardy himself. The British nanny of the Victorian/Edwardian era, it is argued, was a phenomenon which emerged in a society with a particular view of motherhood and childhood (Gathorne-Hardy, 1972: 69). Indeed, the activities of wet-nursing and fostering, as well as the spatial separation of the nursery within the Victorian middle-class home, are used as evidence for 'a tradition of allowing other people to look after one's own children' (Gathorne-Hardy, 1972: 69). Whilst Gathorne-Hardy's arguments appear over-simplistic in comparison to more recent feminist research on motherhood and childhood in the Victorian and Edwardian periods (see, for example, Lewis, 1980), there is little doubt that the so-called 'hey-day' of the British nanny is intricately bound up with the social construction of motherhood by the middle classes of this period. It is the social construction of motherhood, but in a different form, which lies behind the return of the nanny in contemporary Britain, and its importance as a major category of waged domestic labour in this country.

As is widely acknowledged, 'state policy in Britain has long held the mother personally responsible for childcare' (Crompton and Sanderson, 1990: 49). In earlier decades of the twentieth century this view was manifested in various campaigns aimed at improving general infant mortality figures. Educating the mother rather than the provision of alternatives to maternal care or policies aimed at improving the material living conditions of all, was seen as the key to reducing infant mortality (Lewis, 1980). Later, with reduced infant mortality, infant psychology came to be seen as exclusively dependent on the nature of the mother–child relationship. Indeed, the arguments of Bowlby (1958) in particular, and to a lesser extent the popular child psychology works of Spock (1966), proved extremely potent forces in shaping the social construction of motherhood in Britain through the immediate post-war years. Bowlby's arguments were widely interpreted to mean that *any* separation of mother and child was likely to prove damaging, both to the mental health of the child, and to that of the future adult, and proved critical in legitimating the dismantling of nursery facilities established during World War Two (Summerfield, 1984). These same arguments also played a central role in shaping state policy towards formal childcare provision through much of the post-war era. With the dominant ideology requiring that childcare be home-based and provided exclusively by the mother (and not a substitute), the already extremely limited number of local authority day-care places available to young pre-school-age children were set aside increasingly for children meeting exceptional criteria:

Prolonged separation from the mother is detrimental to the child. . . . [W]herever possible the younger pre-school-age child should be at home with his mother and the needs of older pre-school-aged children should be met by attendance at nursery schools or classes. . . . [T]he responsibility of local authorities should continue to be limited to arranging for the day care of children who, from a health point of view, or because of deprived or inadequate backgrounds, have special needs that cannot otherwise be met.

(Ministry of Health Circular 37/68, 1968: 1,
quoted in Moss, 1986:27)

Formal day-care provision for pre-school-age children in Britain through the post-war period therefore has been constructed exclusively for children either with special needs or for those with 'inadequate parents'. Indeed, the only widely available form of collective childcare provision for the majority of children born in post-war Britain has been voluntary sector playgroups (Finch, 1985). Such forms of care have provided a valuable, but essentially supplementary, service to the full-time care of pre-school-age children by mothers. Motherhood, for much of the post-war period in Britain has been constructed as a full-time, home-bound occupation, in which the needs of the mother have been seen as secondary to the needs of the child.

Before the widespread expansion in women's employment, such a situation was relatively uncontroversial. Sustained by an ideology of domesticity and home-making, women's role within society was clearly defined as that of wife/mother/domestic labourer. Employment made things more complicated (see, for example, Brannen and Moss, 1991). As Crompton and Sanderson point out, in meeting the labour force shortages of the 1960s, large numbers of British women found themselves faced by two contrasting and conflicting identities; that of mother and that of citizen (Crompton and Sanderson, 1990: 49). The response of successive British governments to this situation has been limited in the extreme.

As Moss (1986) points out, the views of the recent Thatcher administrations on childcare provision have been clear cut. The correct place for young pre-school-age children through the 1980s was still seen as the parental home; the best care, that provided by the mother (Moss, 1986: 27). Furthermore, whilst working parents were acknowledged to exist, the state response to their needs for childcare provision was governed by broader ideology. Thus, in line with 'rolling back the frontiers of the state' and the promotion of individual/consumer choice, childcare provision for working parents was seen as, and continues to be seen as, 'a private matter, to be left to parents themselves to organise or, in some cases perhaps, to employers in collaboration with employees'

91

(Moss, 1986: 27). Such views are reflected clearly in some of the childcare initiatives unveiled during this period, summarised fully by Mottershead (1988). The Under Fives Initiative (1983), for example, saw the establishment of fifteen voluntary agencies with a £6m budget over a fixed three-year period, and aimed at setting up a variety of projects and/or services for the under-fives. Such activities, however, were seen by central government as 'pump priming', to be taken over by employers on the termination of state funding. Similarly, the conspicuously unsuccessful Childcare Information and Referral Services were designed as local voluntary agencies to provide information for parents on the range and availability (frequently both non-existent) of childcare services in local areas.

Given the above, working parents in contemporary Britain are faced by a situation in which it is virtually impossible to obtain full-time public sector day-care places for young pre-school-age children, and by considerable geographical variation in the provision of places in primary school nurseries (Owens and Moss, 1989; Pinch, 1987). The private sector therefore is the only possible solution for providing care for pre-school-age children if both partners are to remain in full-time work. Such is undoubtedly behind the growth of private nurseries in Britain through the 1980s. It is also behind the expansion in use (and registration) of childminders through this period. But there is no doubt that the rise of the nanny through the same period is also uniquely related to this situation. At one level, the re-emergence of the nanny, like the growing use of childminders and private nurseries in Britain, can be interpreted simply as a private response by parents (and by employment agencies) to the absence of widely available public sector day care. But, at another level, it reflects the dominant ideology of childcare, that is, that home care in the parental home is best.

All this is vitally important to the resurgence of waged domestic labour in Britain generally through the 1980s. The conjuncture between state policy throughout the post-war period regarding the care of pre-school-age children; the expansion of, and potential further growth in, women's employment within the higher-level service class occupations through the 1980s; and such women's tendency to form partnerships with, and have children with, men in a similar class position (Bonney, 1988), means that service class households' potential need for childcare services will frequently expand beyond the nine-to-five, five-days-a-week service provided by private nurseries and most registered childminders. Couple this with the still widespread ideology of the parental home as the best site for childcare, and the conditions in which a maternal substitute form of childcare could emerge (i.e., the form of childcare service provided by the nanny) are readily understandable.

As is evident from our argument thus far, the 1980s in Britain saw the combination of various trends and circumstances, some new (such as the growth of women's employment in the professions), some (such as state policy towards childcare provision) rather more established features of post-war Britain. In advancing this argument we have emphasised the importance of the connections between these developments and the mode of domestic service provision within those middle-class households where both partners are in full-time employment. At least as currently constructed, certain domestic activities, notably child-care (and particularly the care needed by young pre-school-age children) are incompatible with the dual-career pattern of working. Moreover, the continued labour intensive characteristics of other strongly gendered domestic tasks (particularly cleaning and ironing), pose major potential problems for many middle-class women in professional, managerial and administrative occupations. However, at this stage of the argument the problems were left as potential. Although difficult to synchronise, household cleaning and ironing *could* be fitted into middle-class women's non-work time, typically either at weekends or in the evenings. But, if other activities come to impinge on this same 'spare time' then clearly the potential problems become actual ones. This, as we now show, is precisely what happened in Britain in the 1980s.

MIDDLE-CLASS CULTURE AND CONSUMPTION IN THE 1980s

That something 'big' was happening to middle-class lifestyles, consumption patterns and culture in Britain through the 1980s is indisputable. This was the decade of the lifestyle stereotype; the decade of the 'yuppie', the 'dinky' and the 'bobo'. Journalists, such as Leadbetter (1989), referred to the fresh-pasta-eating, French-wine-drinking, southern-based professional middle classes, driving their Audis, BMWs and Golf GTis to holidays in Tuscan villas and aspiring to own small farmhouses in southern France. At the same time academics, for example Thrift (1989), used maps of Laura Ashley and Country Casuals shops, and later estate agent data on the purchase of listed houses, to argue for the emergence of a distinctive south-east 'service class culture', rooted in heritage and the countryside. In short, two elements of middle-class culture – the urbanite and the pseudo-rural, the Barbour jacket/Hunter wellies/Range Rover brigade and the wine bar/French restaurant goers – were being promoted as at the heart of an emergent new middle-class culture in Britain. As Savage *et al.* (1992: 104) have pointed out, such work was far from academically rigorous. In what follows therefore we prefer to draw exclusively on Savage *et al.*'s account, this being the most recent examination of the consumption patterns of middle-class men and women in Britain in the 1980s.

Savage *et al.*'s study of the consumption patterns of the middle classes in Britain in the 1980s is based on 1987/8 market research data. Theoretically their account draws heavily, and inevitably, on the arguments of Bourdieu (1990), and particularly on his concept of cultural 'battles' (themselves grounded in the divisions between economic and cultural capital) between the dominant and subordinate groups within the middle classes (respectively, the industrialists, the new petit bourgeoisie and the intellectuals). For Savage *et al.*, however, unlike others (see, for instance, Urry, 1990; Bagguley *et al.*, 1990), the British middle classes are a qualitatively different phenomenon from the French. Indeed, Savage *et al.* identify three distinctive British middle-class lifestyles and patterns of consumption from their analysis of the market research data, all with distinctive social bases. The first they label 'the ascetic'. These are the standard bearers of the healthy lifestyle, whose consumption patterns score highly for sporting activities such as climbing, yoga and hiking and for high culture performance activities such as classical concerts, theatre and dance. The social base of this group is public sector professionals working in education and welfare – in Bourdieu's terms, a group high in cultural capital but with relatively low economic capital. A second group is the private sector professionals, typically working in law, financial services, personnel, economic advice services, computing and data processing, marketing, advertising, public relations, sales and purchasing. The fascinating feature of this group is that they genuinely appear to sample, in true post-modernist fashion, both the cult of health and body maintenance and the culture of extravagant excess based on heavy drinking and foreign food. Finally, there is the non-distinctive group. Comprising administrative workers in central and local government, '[apart from] their enthusiasm for bowls and their general aversion to squash [they] neither engage in nor avoid any distinctive form of consumption' (Savage *et al.*, 1992: 110). Later (p. 116), the authors argue that this same group is the nearest approximation to Thrift's heritage- and tradition-based 'service class culture'. However, it is their cultural indistinctiveness which matters. Indeed, Savage *et al.* account for this in terms of the way in which their careers and security are bound up with large organisations.

Whilst all of this is fascinating and important, it is where Savage *et al.* move on to examine the mediation of British middle-class consumption patterns in the 1980s by gender that their work becomes of particular importance to our concerns. The clear pattern to emerge from their analysis is that middle-class men engage in a wide array of sports activities, as well as social drinking and eating out. By comparison, middle-class women take part in a more restricted range of sporting activities (typically non-competitive, such as yoga, keep fit and horse riding) and are high attenders at cultural events such as ballet. Savage *et al.* speculate on why this is the case, suggesting initially that such patterns may be the result of a negotiated division of labour between partners (p. 123). In our view this is highly unlikely and we prefer their second suggestion, namely that differential gender patterns of consumption reflect both time constraints on women (specifically childcare) and male exclusionary practices (ibid.).

In spite of the variation in range and type of cultural practices exhibited by middle-class men and women, it is clear from Savage *et al.*'s work that in contemporary Britain middle-class men and women *both* engage in a number of such activities. This is unsurprising. The commodification of culture and leisure in Britain through the 1980s was paralleled by the emergence in middle-class discourse of the concepts of 'leisure time' and 'quality time'. Opposed explicitly to time spent in the activities of paid employment, 'leisure' and 'quality time' have increasingly been constructed by the middle classes as the time for 'self', that is as providing the time–space to engage in activities which men and women both want to do and value doing. Furthermore, as the commodification of cultural and leisure activities proceeded apace through the 1980s, if not the recession-bound 1990s, the material resources existed to sample such activities in abundance. Such might be expected to create major problems regarding those labour intensive domestic tasks referred to earlier.

The first such problem is a straightforward one. With middle-class women, as well as middle-class men, actively engaging in a range of cultural practices, the potential 'spare time' for the gender-segregated tasks of cleaning and ironing becomes very difficult, if not impossible, to find. These time–space difficulties are by no means the only ones however. More important is the construction of time outside employment as 'quality time' and 'leisure time'. Domestic labour, as most feminist research has emphasised, constitutes a clear form of work, albeit conventionally unwaged. Moreover, the labour intensive domestic tasks are hard physical work. Such activities, in contrast to the more pleasurable domestic tasks of childcare activities and cooking, fit uneasily with the concepts of 'quality' and 'leisure time'. The conjuncture between the commodification of culture and leisure and contemporary middle-class discourse in Britain, then, has generated a context in which middle-class households in which both partners are in full-time paid employment face potential conflicts in leisure time use. However, this is not simply a potential conflict between doing *either* household cleaning and ironing *or* a more pleasurable something else. *It is also a potential conflict between partners over who is doing what sort of activity and when.*

The conditions therefore in which demand for another distinctive form of waged domestic labour (the cleaner) could expand are, as with the nanny, contingent ones. But they are very different. In this case it is the conjuncture between the gendering of the labour intensive tasks themselves, the increasing commodification of culture and leisure through the 1980s, the gendered consumption patterns of the middle classes, and middle-class discourse on leisure time which have proved critical. Both the pattern of, and discourse surrounding, cultural practices through the 1980s – as well as the material resources of middle-class households – presented the potential conditions in which middle-class women (on whom the burden of these labour intensive tasks falls) and men, could question, for some – if not all – domestic tasks, the predominantly unwaged tradition of domestic labour in post-war Britain.

Having examined the circumstances and trends which conjoined in the 1980s in Britain to produce the conditions in which middle-class households with both partners in full-time service class occupations could see the need to substitute particular categories of waged domestic labour for their own unwaged labour, it is now time to open up the black box of such households. We turn therefore to examine why individual households of this kind employ the cleaning and nannying forms of waged domestic labour. The necessity of shifting the scale of analysis to the household should be obvious. As we argued above, the potential conditions outlined so far are no more than that. They provide part of the causal context in which we can account for the resurgence in demand for waged domestic labour generally in Britain in the 1980s, and for the expansion in demand for particular categories of waged domestic labour through this period, but it is only middle-class households which can realise this potential.[7] However, it is not just our conceptual position which requires us to move the scale of analysis to that of the middle-class household. As we showed in Chapter Two, our surveys of middle-class households with both partners in full-time service class occupations in Tyne and Wear and Berkshire revealed that in both areas roughly one third of such households were using waged domestic labour. Waged domestic labour, therefore, is not a universal feature within middle-class households in contemporary Britain. Given this, there must be something (or things) particular to certain middle-class households in contemporary Britain which lead them down the route of waged domestic labour substitution in general, and towards using the particular forms of the cleaner and the nanny. If we are to account for the fact that the potential conditions outlined thus far become realised in the employment of cleaners and nannies within middle-class households, we must focus on what is going on within those middle-class households which have substituted these forms of waged domestic labour for their own unwaged labour.

MIDDLE-CLASS HOUSEHOLDS, CLEANERS AND NANNIES

The argument which follows draws extensively on and develops that of other papers (Gregson and Lowe, 1993, 1994). Our starting point here, though, is with Morris's concept of household domestic labour coping strategies (Morris, 1985). Morris argues that, in households in which the female partner is in paid work, partners face a choice between three domestic labour coping strategies. One is that the female partner can perform the domestic work herself, thereby increasing her own work burden considerably. This results in a traditional form of the domestic division of labour, as well as in the so-called 'double day' for women. A second option is for the male partner to assist in domestic labour; a strategy which results in a shared form of the domestic division of labour. Finally, household members can either be assisted by or replaced by another non-household member. In most cases reported so far this 'someone else' is usually a female kin relative – typically the female partner's mother, less frequently a sister (Martin and Roberts, 1984). It is the social relations of kin therefore – of love and of helping out – which underpin this domestic labour coping strategy. However, for middle-class households in service class occupations the kin option is usually unavailable. As already discussed in Chapter Two, although the subject of considerable debate (see, for example, Savage *et al.*, 1988; Savage, 1988; Savage *et al.*, 1992), the spatial mobility of those working in service class occupations, relative to those in manual and non-manual work, is marked. As a result, it is widely recognised that kin relatives are unlikely to be available to such households, at least on a regular basis.[8] Such a situation certainly applied to our middle-class respondent households. Indeed, at the time of our contact with them, only two of the twenty-nine north-east households interviewed had either parents or siblings living in the region, whereas in the south-east only two of the forty interviewed households had kin relatives living in the region. Given this situation, the only 'someone else' option open both to middle-class households in general and to the majority of our specific case study households is/was to substitute waged for unwaged domestic labour.

The three options presented by Morris represent a set of potential ways in which households with a female partner in paid work can satisfy their need for domestic labour. If we simply followed this scheme we could, therefore, just present the use of waged domestic labour by a significant proportion of middle-class households in contemporary Britain as a modified, class-specific version of the unwaged substitution option, one which manifests the pattern of female labour force participation within these households. However, in our view this is a relatively simple (situation-response) way of conceptualising the domestic labour

97

coping strategies of households. Instead, we favour a more contextual approach. Thus, in the first part of this section we argue that the particular form of the domestic division of labour adopted by individual middle-class households, and the partner gender identities which underpin this, is of critical importance to understanding why some middle-class households substitute the cleaning form of waged domestic labour for their own unwaged labour. Following on from this, we show how in certain other middle-class households, it is changing attitudes to the appropriate uses to which middle-class 'leisure time' can be put which lie behind the substitution of the cleaner for unwaged household labour. Beyond this, we consider the rather different instance of the employment of the nanny form of waged domestic labour.

Particular forms of the domestic division of labour, domestic labour coping crises and the cleaner form of waged domestic labour

In another paper we indicate some of the associations which exist between the particular form of the domestic division of labour adopted by partners and the use of specific categories of waged domestic labour (Gregson and Lowe, 1994). Here we develop these arguments to show how the substitution of one particular category of waged domestic labour (the cleaner) for unwaged domestic labour is inherently bound up with the form of the domestic division of labour between partners.

In our work we have employed a sixfold classification of the domestic division of labour between partners. This classification is based on the relative importance of individual domestic labour tasks to household reproduction and on the extent to which these tasks, particularly the strongly gender-segregated labour intensive tasks, are shared between partners (for a full discussion of this, see Gregson and Lowe, 1993). The particular forms of the domestic division of labour between partners which we identify are respectively: the traditional rigid (in which the female partner does all the major regular and minor regular household tasks); the traditional flexible (in which the female partner is assisted by the male partner in a very minor capacity); the conventionally shared (in which both partners share the gender-neutral tasks, but where the female partner continues to perform the labour intensive gender-segregated tasks alone); the predominantly shared (where the gender-neutral plus some of the gender-segregated tasks are shared between partners); the fully shared (where the gender-neutral and a majority of the gender-segregated tasks are shared between partners); and the role reversal (a situation which we regard as similar in nature to the conventionally gendered traditional forms of the domestic division of labour).

When we examine the use of different categories of waged domestic labour by middle-class households with both partners in full-time service class occupations, we find that households with particular forms of the domestic division of labour tend to employ particular categories of waged domestic labour. This is especially so in the case of the employment of a cleaner. Thus, of the 69 households interviewed by us, 26 (38 per cent) exhibited a traditional form of the domestic division of labour prior to employing any form of paid domestic assistance. Of these, 22 (85 per cent) went on to employ a cleaner, either alone or in combination with a nanny. In contrast, only 21 (49 per cent) of the 43 households with shared forms of the domestic division of labour went on to employ a cleaner. More noticeable still is the difference between households with fully shared forms of the domestic division of labour and the remainder with shared forms. Of the 22 households with fully shared forms of the domestic division of labour prior to the employment of paid domestic help (32 per cent of our interviewed households), only 7 (32 per cent) went on to employ a cleaner. Such findings suggest that whereas middle-class households with traditional forms of the domestic division of labour are highly likely to substitute the cleaning form of waged domestic labour for the unwaged labour of the female partner, households with fully shared forms of the domestic division of labour are much more unlikely to do so. Why should this be the case?

At one level we can answer this question in terms of physical time–space and its availability. Our investigations revealed consistently that, whilst traditional forms of the domestic division of labour proved to be workable domestic labour coping strategies for dual-career partnerships without children, the advent of childcare responsibilities – or even pregnancy itself – generated a domestic labour coping crisis for such households. This is well illustrated by the Allen household, who now employ a cleaner and a nanny.[9] John and Carol Allen are both in their late thirties and have three young children, one of whom is primary school age. John is a surgeon; Carol a medical researcher. The particular form of the domestic division of labour between them is traditional. The birth of their first child initally alerted Carol to the difficulties with this domestic labour coping strategy:

CA: It upset me not being able to have the organisation to do all the housework and look after Arabella and keep everything spick and span. That annoyed me. But I tried. I did try.

However, it was the birth of twins which exacerbated the coping crisis further. Whilst on maternity leave:

> everybody decided, including myself, that it would be a good idea if I got some help. I was exhausted really and I couldn't cope with these three and do all the housework, keep him happy and go back to work!

99

Faced with a situation in which 'spare time' had to provide the time–space for all the domestic tasks and childcare activities, the female partners within households with traditional forms of the domestic division of labour found themselves confronted by the classic indivisibility problem. Even with the assistance of all forms of domestic machinery, they could not fit both domestic labour and childcare into the time–space available. In contrast, households with shared forms of the domestic division of labour encountered no such pressures on female labour time. In such circumstances, childcare tasks and activities were simply assimilated within the shared division of labour. Thus, whilst one member of the partnership, say, prepared and cooked the evening meal, the other bathed children, read stories, and so on. Here the Browns, who now have a nanny, describe a typical evening's work. Maria Brown is an interior designer in her mid-thirties. Her husband is a probation officer. They have one pre-school-age child.

MB: In the evenings when we get in we do the jobs. For example, last night we worked in the garden. And then I took her in, bathed her and got her ready for bed, while Peter put everything in the skip. I brought her down and read her a story and, while I did that, he made the dinner. That's the usual pattern.

Whilst there is a basic appealing quality to the time–space argument, there is no doubt that the argument has its explanatory weaknesses. Fundamental amongst these is that, whereas the problems experienced by the female partner in traditional arrangements are immediately obvious from such an argument, it is far from apparent why the resolution of these problems should be the employment of a cleaner. Reverting for a moment back to Morris's three potential options, why is it that such households do not adopt a more shared form of the domestic division of labour? It is at this point that it becomes important to appreciate that the particular form of the domestic division of labour found within individual households is itself an expression of partner gender identities and the place of domestic labour within these.

We can best begin to develop this argument by making reference to three highly contrasting case studies.

The Collinses (now with a cleaner and a nanny)

Andrew and Pauline Collins are both in their early forties. Both have always worked full-time. She is a medical researcher. He is an investment banker. They have two children, one of whom is just primary school age, and did not employ any waged domestic labour before their birth. At this time they organised domestic work on traditional flexible lines. Pauline Collins cited three factors in explaining this pattern of domestic work: her familial background, as well as that of her husband; her 'late' marriage, and social expectation. Neither her mother nor her mother-in-law worked in paid employment:

PC: Their husbands didn't want them to work. They were *proper* [her emphasis] housewives, and I think that you're indoctrinated into their way of thinking.

This pattern of behaviour, although challenged by her own highly successful career in medicine, was also overlaid in Pauline Collins's own mind by what she referred to as her 'late' marriage at 32. Up until then she had always done domestic labour for herself. Marriage saw the merger of all these ideas:

PC: I'd always done things so what was the point in making a fuss about it? Also, when we were first married, I was very keen to be a *proper* wife. When you're newly married you do *so* want to please. So, in a way I kind of slotted into that role.

Thus, whilst she challenged traditional gender expectations in paid work, Pauline Collins chose not to at home. She actively wanted to see herself, and to be seen by her husband, by her family and by her in-laws, both as the successful career woman and the 'proper wife'. In complete contrast to this is the Davies household.

The Davieses (now with a cleaner and a nanny)

Robert and Sarah Davies are both university lecturers. Sarah Davies is in her late thirties; Robert is in his early forties. They have one pre-school-age child. However, unlike the Collins, they employed paid domestic help before the arrival of this child. Before employing any help they organised domestic work on fully shared lines. However, whilst the physical labour in this household was shared, Sarah Davies emphasised that it was she who had taken all the decisions about which tasks were to be done when:

SD: It was a shared physical division of labour, but my tolerance of untidiness and dirt is at a much lower level than his. So I was always the initiator. . . . We always used to fight when it came down to doing it. He'd be saying, 'It looks perfectly all right', and I'd be saying, 'No it doesn't. Look at all those cobwebs! We've got to do something about this.'

Her explanation for this situation stressed cultural forces:

SD: I think it's because if people come in from the outside you're conscious that they think it's women who are responsible for those

101

things. And they'll think, 'God what a slut!' if they're not done. There's more pressure on women. . . . Small bits of fluff become mountains. I [couldn't] settle until they were dealt with. But he could. He could quite happily live in a dung heap and not notice. *He just would not notice!!*

Committed to a fully shared form of the domestic division of labour, Sarah Davies felt that it was she, rather than her husband, who was being judged by family, friends and outsiders. Furthermore, both her own and broader social expectations meant that she could not turn the proverbial 'blind eye' to domestic tasks. Consequently, and given her partner's inability to 'see' domestic labour, she found herself continually pressurising him to do specific tasks. The reproduction of this particular fully shared form of the domestic division of labour therefore was characterised by tension, by inter-partner conflict and by internal conflict for Sarah Davies.

In complete contrast to this again is another household with a fully shared form of the domestic division of labour. This is the Brown household, whom we met earlier.

The Browns

In comparison with the Davies, the most striking thing about the Browns is the consensual way in which the tasks were (and are) shared between the two partners. Earlier, we saw how they coped with the competing time–space demands of childcare activities and domestic tasks. Here we see that this same pattern of accommodation characterises this partnership's total approach to domestic labour:

INTERVIEWER: And things like the ironing. Would you do that?
MB: We do our own. Ironing is the one job which we both particularly hate. Occasionally, Sunday evenings, if I want to iron mine, I'll say, 'Have you got any shirts that need doing for the week?'
INTERVIEWER: And things like the toilet. Who cleans that?
MB: Whoever's doing the bathroom. Or, for instance, Peter'll do the bathroom and I'll do the kitchen.

The instances of the Browns, Collins and Davies demonstrate that the form of the domestic division of labour adopted by partners, as well as being a household domestic labour coping strategy, is also a manifestation of partner's ideas about domestic labour. As we have argued previously, domestic labour, and particularly the traditional form of the domestic division of labour, is intricately interwoven with patriarchal constructions of femininity (Gregson and Lowe, 1993). Within this construct motherhood plays a central, although not exclusive, role. However, it is the characteristics associated with motherhood – caring, nurturing and servicing – and their assumed manifestation in both the gender division of labour and the traditional form of the domestic division of labour, which can play a particularly potent part in shaping the gender identities of individual men and women. As Pauline Collins expressed it, domestic labour, for her, was not just part and parcel of being a wife and mother. Instead, doing the job well was essential to being a 'proper wife', that is to her whole identity as a woman. Much the same ideas were also apparent in Carol Allen's comments earlier, and being unable to accomplish all the domestic tasks meant that she saw herself as failing as a woman.

All this is of fundamental importance to accounting for the reasons why those middle-class households whom we interviewed with traditional forms of the domestic division of labour showed such a clear tendency to employ the cleaning form of waged domestic labour following the onset of childcare responsibilities. For them changing the form of the domestic division of labour to one involving sharing between partners was simply not an option. To do this would have required a complete restructuring of the relationship of these male partners to domestic labour, as well as a fundamental reappraisal of their identities as men. Furthermore, for many of the women in these households, sharing domestic labour would have been equally difficult, for – as we have seen – domestic labour is centrally bound up with their concept of themselves as women. In one sense then, we can see that the employment of a cleaner by respondents with traditional forms of the domestic division of labour is closely bound up with both partner's gender identities. However, it is to the female partner, undoubtedly, that this form of substitution has most to offer. A clean house, as we have seen, is important to many of them. This is because living in dirt and grime signifies not just domestic mess, but failing to be the 'proper wife'. And yet the contradiction which the onset of childcare responsibilities brought for these women was that they began to fail in part of their role as the 'proper wife' at precisely the same time as they began to engage in other valued domestic activities, notably those associated with motherhood. Employing a cleaner is the only way in which the women in these partnerships could easily resolve this basic, undermining and guilt-ridden contradiction.

The situation presented by the Davieses is very different, although once more it is partner gender identities, and their relationship to both the gender division of labour and the traditional form of the domestic division of labour, which are well to the fore. This household represented one of a minority of cases which we encountered in which households with fully shared forms of the domestic division of labour substituted the cleaning form of waged domestic labour for unwaged household labour. The reasons behind this substitution were complex. Neither Sarah Davies, nor her partner, saw domestic labour as a positive part of their gender identities. Indeed, if anything, the reverse was the case; certainly for Sarah Davies, who articulated a feminist position in many of her comments. Thus, the domestic division of labour worked out by this partnership could only be one of the shared variants, but one achieved only at considerable cost. Although Robert Davies was not opposed to doing domestic tasks, Sarah Davies found that it was continually she who had to push when specific tasks needed doing. Regardless of later rationalisation and intellectualising, the differential part played by domestic labour in the socialisation of women and men had created a situation in which this particular woman 'saw' domestic labour, whilst her partner did not. We encountered this situation time and time again in households with shared forms of the domestic division of labour. However, whilst this point is highly significant, it is the way in which partners respond to it which is critical in determining whether or not households with fully shared forms of the domestic division of labour eventually substitute the cleaner form of waged domestic labour.

In the case of the Davieses, the response was to argue over the tasks. This can only be understood in relation to the internal conflict experienced by Sarah Davies. Whilst she rejected the association between women and domestic labour at a personal level, she felt that it was she, and not her partner, who would be criticised if domestic tasks remained either undone or poorly done. Such social censure was something which she felt to emanate from her mother and from outsiders generally. Given such pressures, domestic tasks could not remain undone. But, at the same time, the form of the domestic division of labour meant that Sarah Davies was invariably 'getting at' her partner. Ultimately, as she says:

> We decided that it was ridiculous to be arguing over cleaning. It just wasn't worth it. We were spending more time arguing over how clean we wanted the house and the five degrees of difference between my standards and Robert's standards, that it was a nonsense. So we had a cleaner. . . . She came on a Friday. The house was clean for the weekend. That was when we saw it. For the rest of the time it didn't matter and that was OK.

For this household, employing a cleaner represented the only domestic labour coping strategy capable of eliminating partner conflict over domestic labour.

In comparison, the Browns illustrate the majority situation for households with fully shared forms of the domestic division of labour. In these cases, whilst there is no doubt that the same differences in 'seeing' domestic labour operated, the response was very different. Instead of arguing about it, both partners in these households simply got on with the work. We have found it difficult to get to the bottom of this particular arrangement. However, after much deliberation we consider that the thing which sets them apart from the Davieses is their pragmatic, as opposed to ideologically loaded, approach to domestic labour. The Browns approach domestic tasks as 'jobs', indeed as necessary jobs. There is an acceptance here, and whilst neither partner actively identified with domestic tasks, in the manner of the women with traditional forms of the domestic division of labour, they clearly enjoyed working on domestic tasks as a partnership. Much the same sentiments were expressed by other partners with similar arrangements, as for example, the Evans (now with a nanny). Richard and Elaine Evans are both retail managers working in a large city centre department store. They have one pre-school-age child and did not employ any paid domestic help before her birth.

EE:	We'd just moved into this house then and I suppose there was quite a lot of decorating. I think it's fair to say that I did the majority of the cooking. We go together to do the shopping. I make a list and Richard pushes the trolley. When it comes to the cleaning, I'll do the downstairs, Richard'll do the upstairs. Richard will do the ironing, fill and empty the washing machine. It's a pretty fair division.
RE:	I suppose there's more of a division of the kitchen activities. I'll tend to wash up and wipe up. Elaine'll tend to cook. I'll do the windows.
EE:	The thing is I don't think either of us dislike any of the domestic chores.
RE:	That's right. For example, I enjoy cleaning the bathroom. I suspect that it's a pretty fair division, although I think that Elaine'll tend to see more. And I think that that's partly education, partly the person. I might, by the time the evening came around, nine to nine-thirty say – I'll say, 'Right, that's it. Let's jack it in.' And she'll say, 'No, there's more to be done yet.'
EE:	Well there's a difference isn't there between a man and a woman. For example, I'll say, 'These table cloths need doing', or 'We need to take these cushion covers off', and 'We need to change the sheets on the bed.'
RE:	Or the kitchen sink and that sort of thing.
INTERVIEWER:	And what about DIY and that sort of thing?
RE:	We do that sort of thing together and Elaine is as expert as I am in decorating terms. When it comes to rebuilding

or making things, then that's me, because I know how the various tools are used.

INTERVIEWER: So when you first set up house together, did you actually discuss this?

EE: No! We fell into it! I had lived by myself in the flat for eight years and I was pretty independent. I knew how I liked to run things. And Richard had had his own house so he knew how he liked to run things. I don't suppose we had wildly different ideas, but we were both pulling in the same direction, although we might do it slightly differently. And so, basically, we were probably very similar in terms of what we wanted to achieve and how we did it.

RE: It wasn't difficult. What was difficult and what we've learnt is how to do things the way the other person likes them to be done. And that's a compromise. Whether it's painting brush strokes on a wall – there's this way of doing it and there's this way of doing it, and it's a matter of reluctantly agreeing that the other's way of doing it might be better . . .

In contrast, for the intellectual partnership of the Davieses, the analytical significance of domestic labour was never far away. From this we conclude that whilst households with fully shared forms of the domestic division of labour are less likely in general to substitute the cleaner form of waged domestic labour for the unwaged labour of partners than those households with traditional forms of the domestic division of labour, both the ideas and identities of partners can mediate this tendency. Partnerships which adopt a purely pragmatic approach to domestic labour are those for whom sharing proves to be a robust and consensual form of household domestic labour coping strategy. In contrast, for those households which recognise explicitly the whole gamut of con-notations surrounding domestic labour and its organisation, sharing is often a conflict-strewn path. For households in the latter category, there is little doubt that waged labour substitution in the form of a cleaner is a more amicable form of domestic labour coping strategy, than that involving the unwaged labour of both partners.

Thus far in our explanation of why individual middle-class house-holds substitute the cleaner form of waged domestic labour for their own unwaged labour, we have stressed only issues internal to domestic labour itself. There is, however, another dimension to this. Whilst, for the most part, it is issues internal to domestic labour and its organisation within households which push individual households to use the cleaner form of waged domestic labour, pressures emanating from other, non-domestic sources, can produce the same result. In our investigations this most consistent 'other' was competing uses of partner's 'leisure time'.

Quality time and the cleaner . . . or 'there was no way that we were going to spend our time together doing that!!'

The most important point to establish at the outset here is that the only households we encountered which articulated this position were those with shared forms of the domestic division of labour. This we feel to be significant. Referring back to the previous arguments, it is evident that the need to perform domestic tasks posed no problems for many middle-class women. In such circumstances there is no conflict between the tasks and activities which an individual woman is having to do and those which she values doing. In contrast, for those partnerships with competing leisure-time interests and shared forms of the domestic division of labour, domestic labour, and particularly the labour intensive tasks, pose major problems. It was within such households that arguments concerning quality time were repeatedly voiced.

The following extracts provide some indication of the types of arguments being made here:

The Frazers (now employing a cleaner and a nanny)

Mike Frazer works in local government as a manager. Ruth Frazer is a TV journalist. They have two pre-school-age children and employed paid domestic help before the birth of their first child. The form of the domestic division of labour within this household was then predominantly shared.

INTERVIEWER: So how long did this situation continue for?

RF: I suppose years! I suppose about four or five years! We were both working full-time. And then we both developed definite interests – hobbies. It became, '*This is how I want to spend my time. I want to do this*' [her emphasis]. Time became more valuable. And Mike said, 'Let's get someone in to do the cleaning.' And I said, 'No. We're two adults. If we can't keep this place clean it's a pretty poor show.' And this went on for a couple of months. And by then I was virtually convinced. Because Mike was saying, 'If you rationalise it what you're doing is not buying a slave. You're buying time. Because you are paying someone else to do these jobs; jobs which you'd have to do otherwise. So, if you want to have the weekend free to do what you want to do then you're going to have

107

to find somebody else to do the dirty work.' And that's what we did.

The Greens (now with a cleaner)

Alan Green is a teacher and Linda Green a university administrator. This couple do not have any children and organised domestic labour in a fully shared way.

LG: I was then working at the X centre in Y, and because we worked a seven-day week I had to work one day of the weekend. So there was only one day of the weekend when Alan and I could be together at home. And there was *no way* [her emphasis] that we were going to spend that time cleaning and ironing. So we decided that a cleaner was a good idea.

The Harrises (now with a cleaner and a nanny)

Ian Harris is an architect and Anne Harris a public relations consultant. They have one pre-school-age child and employed paid domestic help before the birth of this child. They organised domestic work on predominantly shared lines.

INTERVIEWER: When did you first employ any help?

AH: That would have been about four years ago (i.e., before the birth of the child).

INTERVIEWER: And that was a cleaner?

AH: Yes.

INTERVIEWER: So what brought about the change?

AH: It started to be that I was working longer and longer hours. And it meant a certain amount of work over the weekend. So we just thought that we needed time to switch off and just do things together. I was also travelling away from home quite a lot. So the time together became more important. And we wanted to do better things with it than brandishing a hoover!!

Two points are immediately apparent from all three of these extracts. One is that faced by time–space pressures on non-work time, imposed either by non-standard working patterns or the development of new hobbies and/or interests, these partnerships were forced to re-evaluate their use of non-work time. In this re-evaluation it was the labour intensive domestic tasks, particularly cleaning, which came out at the bottom of the pile of possible activities to fill this time–space. Although widely seen as important and necessary tasks by these partnerships, as an activity cleaning was seen as 'dirty' and as 'work'. Both terms, along with the discourse of 'quality time' enabled partners to categorise it as an activity in conflict with their ideas about how they ought to be spending their hours beyond paid employment. The second point follows on from this. This is that for these households the employment of a cleaner represents not just a household domestic labour coping strategy, but an enabling strategy which goes beyond domestic labour itself. In the cases considered previously, the substitution of the cleaner form of waged domestic labour for unwaged household labour in one set of cases (the traditional) merely enabled the female partner to perform other competing forms of domestic labour (notably childcare activities). In the other set of cases (the fully shared) it removed a major source of conflict between partners. In this set of cases it enabled partners to do something else altogether.

To summarise: our investigations suggest three fundamentally different reasons why middle-class households with both partners in full-time service class employment in contemporary Britain substitute the cleaner category of waged domestic labour for unwaged household labour. In order of apparent frequency of occurrence these were:

1 To enable the female partner in a household to do other reproductive tasks in the time–space available to her. In this situation households use their material resources to resolve a basic contradiction for the female partner over domestic labour.
2 To enable both partners to do other non-domestic activities. In this situation households use their material resources to pass on certain reproductive tasks to someone else, leaving them 'free' to do other, usually consumption-based, activities.
3 To enable partners to resolve conflicts over domestic labour tasks; particularly over when and how these tasks should be done, rather than over who should be doing them. In this situation, whilst households use their material resources to resolve conflict, the end result is much the same as in (2).

In situations (2) and (3), middle-class households are using their material resources to purchase the labour time of someone else, so as to buy in turn their own leisure time. By comparison, in situation (1), material resources are again being used to purchase the labour time of another person, but this time to free up the labour time of the female partner for other domestic activities and tasks. We consider such findings to be of critical importance to contemporary debate on domestic labour. However, at this stage we do no more than highlight two particular issues. The first is that it is clear that middle-class households have a definite hierarchy of domestic tasks. At the top are childcare activities; at the bottom is cleaning. And what is at the bottom is being hived off by a significant proportion of middle-class households in contemporary

Britain. Secondly, whilst the employment of a cleaner represents a domestic labour coping strategy on the part of middle-class households, at the same time, these actions will have unintended consequences. One of these is that household cleaning in middle-class households is no longer just a gender-segregated task. As we show in Chapter Five, it is also being constructed as an occupation for working-class women. The corollary of this position is that in certain middle-class households cleaning is no longer being seen as a suitable use of middle-class women's time–space.

Having examined why individual middle-class households in contemporary Britain substitute the cleaner form of waged domestic labour for unwaged household labour, we are now in a position to relate these arguments back to those of the previous section. Clearly our earlier argument that it is the labour intensive domestic tasks, rather than domestic labour generally, which pose major problems for these households has been shown to be correct; as has our argument that it is for women that these problems are particularly acute. But, these arguments need qualification. At the moment – although, as we will show in Part Two, the reality is somewhat different – it looks as though, of the two potentially problematical tasks identified earlier (cleaning and ironing), only cleaning is sufficiently problematical to suggest the substitution of waged for unwaged labour. Moreover, as we have seen, whilst pressure on time–space certainly comes into the decision to substitute, for certain middle-class households it is the form of the domestic division of labour which appears to play a major part in leading households down the substitution route. This is not something which could have been anticipated and, as such, pays ample testimony to the importance of the middle-class household in accounting for both the resurgence in demand for waged domestic labour generally, and in accounting for the expansion in demand for particular categories of waged domestic labour in Britain through the 1980s. A second point to emerge is, as we anticipated, the connection between demand for cleaners on the one hand, and, on the other, the interplay between work, consumption and domestic labour. As we saw above, this is a particularly potent combination of forces. Indeed, it creates the conditions in which middle-class households can (as some clearly do) question the need for all domestic tasks to be unwaged. The substitution of the cleaner form of waged domestic labour for the unwaged labour of partners is here related to partner choice, and specifically to value judgements about cleaning, both as a form of domestic labour and as a form of partner activity. That these partnerships can do the labour intensive cleaning tasks is not in question; they have done them and they have the time–space available to do them, if they wish. However, the choice which a significant number of middle-class households are apparently now making is to do them no longer. This choice is bound up with a series of judgements about the quality of leisure time and the perceived incompatibility of cleaning tasks with this.

Finally, it is necessary to relate these arguments to the broader concerns of this chapter, namely the resurgence in demand generally for waged domestic labour in Britain through the 1980s, and the importance of the cleaner category within this. Of the three specific reasons which we have isolated as behind the substitution of the cleaner form of waged domestic labour within middle-class households, only the second – to enable partners to engage in other non-domestic forms of activity – is something which we can pin-point as particular to the 1980s. Varying forms of the domestic division of labour and conflicts between partners are issues which have been around somewhat longer. This would appear to suggest that the demand for cleaners by the middle classes in Britain through the 1980s was strongly bound up with the discourse of 'quality time', middle-class consumption patterns and the incompatibility of cleaning with these. Whilst this is undeniably a major aspect of the expansion in demand for cleaners through the 1980s, its importance should not be over-estimated. Equally significant is the pattern of female participation within the service class occupations, for it is this which lies behind the domestic labour coping crises. The demand for cleaners by middle-class households in Britain through the 1980s, then, is to be accounted for partly by re-evaluations of how such partnerships spend their time outside of employment and partly by a crisis in middle-class household domestic labour coping strategies. For those middle-class households whose domestic division of labour remains traditional, the employment of a cleaner represents possibly the only means of restructuring the domestic division of labour to cope with both the demands of full-time employment and those of motherhood.

Having examined the reasons why individual middle-class households substitute the cleaner form of waged domestic labour for unwaged household labour, we now move on to consider the rather different circumstances surrounding the employment of nannies by middle-class households.

Mother substitutes, the practicalities of parenting for middle-class households and the nanny as a contemporary form of waged domestic labour

It is important to be clear that the form of substitution which takes place when middle-class households employ the nanny category of waged domestic labour is very different to that which occurs with the cleaning category. When middle-class households in full-time service class occupations employ a cleaner they are substituting waged labour for a set of tasks which they have usually done themselves and over a period of years. In the case of the nanny, they are substituting waged labour for tasks which have usually only been performed by the female partner on a full-time basis, and then only for a short duration, i.e., for the period of maternity leave after the birth of the child. This difference is important. As we have shown, in a number of cases the cleaner category of waged domestic labour is employed when it is still physically possible for partners to perform these domestic tasks. The pattern of labour force participation in service class households, however, is such that a full-time form of childcare provision *has* to be found, if these partnerships are to reproduce the dual-career pattern of working. But, having said this, and again unlike cleaning, childcare (both in terms of activities and tasks) is something which partners themselves continue to perform in an unwaged capacity in the hours beyond employment. Beyond this, the two forms of substitution exhibit a second central distinction. In the case of the cleaner, substitution relates exclusively to physical, and frequently hard, dirty, labour. In the case of the nanny, whilst the labour certainly can be physical, it is predominantly construed as care-related. Moreover, whilst the one is related to material phenomena, the other is connected to flesh and blood. Whilst the cleaner is entrusted with valuables and/or objects of great sentimental importance, the nanny is entrusted with the care of what is, almost inevitably, the most valued element of these couples' lives. This, as we show in Part Two, has major implications for the social relations which characterise the two forms of substitution.

It is also possible to identify a third fundamental difference between the use of cleaners and nannies. In the case of the former, middle-class households have no choice but to employ someone else to come into their own homes. Cleaning tasks – unlike say washing and ironing – simply cannot be separated from the homes and the people who generate them. In this sense we disagree fundamentally with Watson (1991) where she argues that domestic labour is being transferred out of middle-class homes into the homes of the working-classes. By contrast, the employment of waged domestic labour for childcare poses a choice for partners. Pre-school-age children, although dependent on their parents, are spatially mobile. Thus, middle-class households in contemporary Britain with both partners in full-time service class occupations face a decision whether to use a nanny (care in the parental home) or a childminder (care in the minder's home). Our investigations shed considerable light on why certain middle-class households opt for the former.

As with our discussion of the employment of the cleaner category of waged domestic labour within middle-class households, we begin our analysis of why such households employ nannies with the comments of some of our respondents. To maintain some sense of continuity, as well as to highlight the differences between the conditions which result in the employment of a cleaner and those which give rise to the employment of a nanny, we have tried – in so far as this is possible – to use as illustration some of the households which we introduced above.

The Browns

MB: We'd never particularly thought about having children. And then we decided that 'the future had arrived'. Friends were having children so we decided that, yes, we were settled and that we had surplus income. We were starting to buy clothes, which we'd never done. So we decided to have a child. And while I was pregnant I was working full-time. And at ante-natal one of the health visitors said, 'What are you going to do afterwards?' And I said, 'I've friends who have childminders, maybe I'll find a childminder.' I didn't know anybody. And to be honest I was a bit wary because you always hear stories. So she said, 'I'll give you a list of childminders, but I'll also give you the telephone numbers of girls that have been to ante-natal but have got nannies.' So she did that. And somebody rang me up and explained that she had a nanny and how it all worked.

INTERVIEWER: So had you any ideas what it was about?

MB: Well just the usual stereotype. That you had to be very wealthy and that she had to live in a flat attached to the house.

INTERVIEWER: So you started to look into this?

MB: Yes. I said to the health visitor that I'd look into it but that we couldn't afford to have anybody living in. And she said, 'You don't have to have anybody living in. Most of the nannies live at home and come daily.' So I thought about it. And the more I thought about it – we're pretty disorganised! And the more I thought about it – the prospect of getting the child out – especially in the winter time, unless you could find someone very close. So it started to sound very attractive.

The Davieses

INTERVIEWER: And then you had Sylvie. Was the decision always that you would have a nanny?

SD: We didn't want a nanny. We never even considered

113

having a nanny. Nannies were things which rich people had. . . . I wanted her to go to a crèche. My feeling was that, if I wasn't going to be looking after her, the important thing was that she had stability and that she knew what the routine was every day, and that she could depend on that. So I tried to find crèche facilities first of all in B. Which was impossible. The only guaranteed childcare that I could get in a crèche was three mornings a week in the university crèche. That was all. That wasn't during vacations, school half-terms and so on. It was absolutely impossible. So then I thought, childminders. But I was much less happy about that because I wanted her to be with a lot of other children, and I didn't particularly like the idea of her going into somebody else's home with maybe just one or two other children. But it seemed that really there was no other choice. So we made some enquiries about childminders and that turned out to be not very satisfactory because they said that they didn't want children in the summer holidays when their own children were back from school. Really it was very strongly related to when the school terms were. So that wasn't suitable. And then somebody suggested a nanny to us. And then I said first of all that that was out of the question. Partly financially and partly because I thought it smacked of elitism – the child being left at home all day and being looked after by one person. For me the important thing was that she was with other children. But there really was no alternative.

The Isards

Julia Isard is a public relations executive. Her husband is a public relations director. They have one pre-school-age child and currently employ both a cleaner and a nanny.

JI: I actually wanted a childminder. I'd got this idea of a nanny as someone with a uniform and a starched apron – the idle rich! So really I thought it would have to be a full-time childminder. I went about asking round and so forth, and I had to find somebody quickly, within two weeks. There were only two names that they could give me within the area and I didn't want him to have to travel more than ten minutes. One I found, and by the time I got hold of her she'd been booked by somebody else. And the other one, I just didn't like her. Also, I found out that she was quite expensive.

When I worked out how much she would cost, it was far more than I had bargained for. So then I got in touch with [an agency] and asked them about mothers' helps. And they said that you had to be in the house with them. So then I said that I didn't think that I could afford a nanny. And then they told me what the price was – and they were less than childminders!! That, plus the convenience of being in his own home. I wouldn't have to flog about packing him up at the beginning and end of each day and so forth.

The Joneses

Janet Jones is a TV producer. Peter Jones is a self-employed film director. They have one pre-school-age child and employ both a nanny and a cleaner.

INTERVIEWER:	There was no considering a childminder?
JJ:	No. Luckily we earn enough money not to have to use a childminder. . . . I wanted her to be in her own home as much as possible. I had to go back to work when she was four months old, and I didn't want her going to a childminder.
INTERVIEWER:	Did you feel that she would be happier here?
JJ:	There were two parts to it. One was that I wanted somebody exclusive to look after her. I felt that she would do better. Not that I'm knocking childminders, because I know that there are a lot of good childminders. But she was so young. And the other situation was that Peter had just gone freelance in his career, and for his business it was important to have somebody here to answer the phone.

The Keiths

Marianne Keith is a TV researcher. Her husband is an architect. They have one pre-school-age child and employ both a cleaner and a nanny.

INTERVIEWER:	So why did you choose a nanny versus, say, a childminder?
MK:	We thought that it was the next best thing to me staying at home.
INTERIVEWER:	Why was that?
MK:	One-to-one attention. And I felt that both for my sake and the baby's that it was so much better for her to be at home. From what I've heard about childminders, from people in a similar situation to mine, they've chopped and changed. Suddenly they don't want to do it any more. Or

they move. I know that can happen with nannies. But I think that if a child is in its own home it's probably more secure and settled. And I wanted Charlotte to be at home in our house, not uprooted in the morning, dumped somewhere and then dragged back for tea and bed.

INTERVIEWER: Did your work demands come into it?

MK: Yes. I was a production assistant at the time. As a PA you can be scheduled on till midnight if you're on a big outside broadcast. And if you're on the local news magazine programme you don't finish till 7, which was another reason why I wanted somebody in the house. I thought, if I've got these flexible hours, and I've got somebody in my own house, I can expect them to be flexible, much more than a childminder, who'll probably have their own kids coming in from school, or husband's demands on them.

In the case of the Browns, the decision to employ a nanny was a gradual one. The image held of the nanny was one which neither accorded with this couple's financial circumstances nor with their living arrangements (a modest three-bedroom semi, albeit in a good area). Indeed, it is the ante-natal worker who made the initial suggestion of a nanny; who indicated that a nanny can be a day nanny; and who put Maria Brown in contact with nanny employers. This couple's decision about the form of childcare provision to adopt, therefore, was one shaped strongly by the ideas of a healthcare professional. Though we can only speculate about this, we are convinced that it is the 'home care is best' argument which underpinned such suggestions. None the less, as is apparent from Maria Brown's remarks, we can see that it is the practical advantages which home-based care appeared to offer to parents (rather than to the child) which this couple found most persuasive in opting for a nanny.

The themes of 'home care is best' and practical advantages also feature strongly in the cases of the Joneses and the Keiths, but in these instances it is the parents who made the arguments. The Keiths are one of the many households which we encountered in which partners articulated explicitly the dominant ideology surrounding childcare. They considered care based in the parental home to provide emotional security for the child and to be the next best form of care to that provided by the child's natural mother. Furthermore, they also saw the one-to-one attention offered by a nanny as important. However, the practical advantages of the nanny for parents were also stressed. Thus, for Marianne Keith, a nanny appeared to offer a more flexible form of childcare provision than a childminder. Similar arguments were made by Janet Jones, who also expressed practical reasons for wanting their child to be looked after in their own home – they needed someone else to be there to answer the phone! In this household, too, we encountered strong negative feelings, and statements about the potential quality of care offered by childminders. For the Joneses, a nanny appeared to offer a more exclusive form of childcare provision.

Somewhat removed from these three instances are the Isards. In this case, as with the Browns, both partners had the impression that a nanny was 'not for the likes of them'. In particular, this couple felt that this form of provision was neither appropriate to their occupational class, nor the most appropriate form of childcare provision. Unlike the Browns, Joneses and Keiths, Julia Isard actively went down the route of trying to get a childminder. However, she found this to be both more difficult and more expensive than she had anticipated. Desperation led her to enquire first about mothers' helps (a non-option for this household given Julia Isard's pattern of labour force participation) and then nannies. As we can see, cost proved a major attraction for this household (as it did to others, particularly those with two pre-school-age children).[10] But, so too did the type of organisational advantages mentioned by the Browns. A nanny was seen to mean less work for the mother at the beginning and end of each day.

Finally, there are the Davieses. Although certain of Sarah Davies's comments echo those made by other households, it was one of a small minority which we encountered. Earlier we stressed that this partnership comprised two academics. We also emphasised that Sarah Davies articulated a feminist position in many of her comments, and this is apparent in her outline of why this household employed a nanny. Unlike any of the households referred to previously, the Davies wished their child to be cared for in collective day-care facilities, deeming social mixing to be imperative. Clearly, then, this household rejected wholeheartedly the dominant ideology of childcare. However, in attempting to find an acceptable form of care, they found themselves confronting the proverbial 'brick wall'. Part-time provision was the best possibility but totally inadequate for their needs. Their fall-back position was to try to find a childminder. But again this did not provide the degree of full-time care required by the Davieses.[11] Finally, and reluctantly, this household was forced to employ a nanny. As with the Browns and the Isards, this partnership felt initially that employing a nanny was beyond their financial resources. In addition it was also seen to offer completely the opposite form of care to that which they wished for their child. One-to-one attention was here interpreted as elitist (compare the Joneses and Keiths who actively sought an exclusive form of care). For this household, the employment of a nanny has to be seen as a case of 'there is no alternative'.

The above case studies suggest that there are three fundamental reasons why individual middle-class households with both partners in full-time employment in service class occupations opt for the nanny category of waged domestic labour, rather than for a childminder. In order of importance, these are:

1 Because the nanny is the only form of childcare provision to offer *care in the parental home*. As such it is the only form of care to be remotely in accordance with the type of thinking which has dominated debate on childcare provision in Britain in the post-war period.
2 Because the nanny offers major practical advantages for parents, including flexibility.
3 Because, as at least some of our respondent households found, the nanny represents a cheaper form of childcare provision than that of a childminder, particularly for those households with more than one pre-school-age child.

Of the three reasons identified, it was undoubtedly the first which proved most critical in pushing our respondent households in the direction of a nanny, rather than towards a childminder. Thus, the nanny was seen to offer a form of care which closely approximated that traditionally provided by a child's natural mother. Indeed, many of our respondents either referred to the nanny explicitly as a 'mother substitute' or implied in their comments that they viewed the form of care provided by the nanny in this light (see, for example, the Joneses and the Keiths above). In making their choice of childcare provision, our respondents saw the nanny as offering 'the next best thing' to what they saw as the type and the quality of care provided by the full-time mother. These comments regarding the nanny as mother substitute are fascinating, not just for what they tell us about general perceptions of the nanny form of waged domestic labour, but for the contradictions which they reveal. We explore these in greater depth in Chapter Six. However, at this stage in the development of our argument, one such contradiction is worth highlighting. This is the mismatch between, on the one hand, the individual constructions of the relationship between motherhood and childcare by our respondent households and, on the other, the general ideas which underpinned their move to employ a nanny. In terms of the former, the women in the households we interviewed had all rejected (albeit for different reasons) the conventional association in contemporary Britain between motherhood and the full-time care of young, pre-school-age children. But, at the same time, for many of these households it was this same conventional association which was used to choose the nannying form of childcare provision for their children. At one level, these households effectively rejected the dominant ideology of childcare, and yet at another they were constructing personal solutions to childcare provision on lines which closely approximated with dominant social conventions.

Whilst there is little doubt that arguments concerning both the nanny as a mother substitute and the nanny as 'the next best thing' to the mother were of primary importance for most of the households interviewed by us, the apparent practical advantages offered by the nanny were also influential. At this point we isolate only those practical advantages which the employment of a nanny appeared to offer households, and not those which they found this form of care to offer in practice. There were three sets of comments which households made repeatedly to us. Firstly, a nanny was considered to offer major *organisational* advantages, in that the child did not have to be moved out of his/her own home in the morning and collected in the evening. This was widely seen to mean less work for parents. Employing a nanny meant that they neither had to organise the child (and his/her own things for the day) nor re-organise themselves to accommodate dropping off and picking up arrangements. A second advantage identified was *flexibility*. Before employing a nanny, many of these households felt that this form of care offered greater flexibility in terms of hours than that provided by a childminder. Thus, it was widely assumed, for example, that it would be possible for parents to ring up and expect a nanny to stay late until they arrived home from work. This expectation was closely interwoven with both the location of care in the parental home (as opposed to in someone else's) and with the assumption that a childminder – but not a nanny – would have competing sets of demands (husbands and children). Finally, our respondent households undoubtedly recognised the potential advantages of having someone else based in their own home for much of the day. Effectively, they were aware that a nanny could provide the form of support service conventionally associated with the *housewife*.

Having examined the reasons why our respondent households employed a nanny in relation to their childcare needs, and why they did so rather than employ a childminder, we are now able to relate these findings back to the comments of the previous section. There we emphasised that the conjuncture between middle-class patterns of labour force participation in Britain in the 1980s, and the dominant ideology of childcare through the post-war period created the potential conditions for the return of the nanny within the British middle classes. Such arguments have been entirely borne out by our investigations of middle-class households with both partners in full-time service class employment. However, at that stage of our analysis we neither anticipated the additional importance of the assumed practicalities offered by the nanny in shaping the decisions of our respondent households, nor the strength of negative comments on childminders. Given the emphasis which many of our nanny-employing households placed on the assumed practical advantages offered by the nanny, we suggest the nanny – much like the cleaner – needs to be seen as a form of household domestic labour coping strategy. But it is clearly an enabling strategy too; one which facilitates the combination of parenthood with full-time service class employment. At one level, the increasing demand for nannies in Britain through the 1980s needs to be seen as indicative of a recognition on the part of a growing number of middle-class households that the nanny offers one of a very limited number of means of negotiating and reproducing the combination of the dual-career pattern of working and parenthood in contemporary Britain. At another level this same increase in demand needs to be read as a manifestation of the way in which the decisions of such households regarding childcare are themselves structured by the dominant ideology surrounding childcare in contemporary Britain. Indeed, it is the acceptance of the 'home care is best' line which enables these households to reject, on the grounds of quality and form, the care offered by a childminder, in favour of employing a nanny. In a sense, then, whilst the employment of the nanny form of waged domestic labour enables middle-class households with both partners in full-time service class occupations to combine paid work and parenthood, such employment is, at the same time, reproductive of very traditional ideas about the care of young, pre-school-age children.

CONCLUSIONS

The foregoing enables us to see why the middle-class households we interviewed substituted the cleaner and nanny forms of waged domestic labour for their own household labour. It has also shown how the potential conditions outlined initially in this chapter were realised within households we interviewed. Thus, we have shown how both the particular form of the domestic division of labour within certain middle-class households and re-evaluations of the place of domestic labour within 'leisure time' led them to substitute a cleaner for either the unwaged labour of the female partner or of both partners. In relation to the nanny we have demonstrated the potency of conventional assumptions and associations regarding childcare and motherhood in shaping household decisions over childcare, as well as the importance of certain assumed practical advantages which the nanny was considered to offer service class households. In both cases, the substitution decisions made by our respondent households were ones made within the parameters of individual households. Moreover, these decisions were shaped both by general ideological debate and by the emergent cultural practices of the new middle classes of 1980s Britain. Absent entirely from this was any specific local area influence. Geography – in the form of local, place-specific practices – had, apparently no influence on our respondent households' decisions to employ either a cleaner or a nanny, or both. Rather than conspicuously adopting the social practices of others – in particular, those of friends and neighbours – the households we interviewed all emphasised the importance of either personal or inter-personal decisions taken within individual middle-class households. Furthermore, in a minority of cases it was suggested to us that the employment of waged domestic labour (particularly of a cleaner, but also of a nanny) was something which households even went so far as to keep quiet about.[12] Such findings provide strong support for Morris's arguments concerning space, place and household dynamics.

In the following chapter we move on to account for how the demand for waged domestic labour by middle-class households has been satisfied. Here we are concerned with establishing the conditions which have made possible the generation of a labour force of nannies and cleaners.

5

UNPACKING SUPPLY

In this chapter we focus on the conditions which made possible the resurgence of, and expansion in, particular waged domestic labour forces in Britain in the 1980s. The explanatory form of our argument is much as in the previous chapter, emphasising the conjuncture of specific trends and circumstances. Moreover, our explanation again concentrates on specific categories of waged domestic labour, the cleaner and the nanny. However, before focusing specifically on these two labour forces it is important to consider why it is inappropriate to refer to a pool of waged domestic labour in contemporary Britain, and to outline the relation between demand for waged domestic labour and labour supply in Britain through the 1980s.

WAGED DOMESTIC LABOUR FORCE CHARACTERISTICS IN CONTEMPORARY BRITAIN AND THEIR MEDIATION BY CLASS AND THE LIFE-CYCLE

At first sight it might seem odd to argue that it is inappropriate to refer to a general pool of waged domestic labour in Britain and wrong to think in terms of a set of conditions which have produced a general waged domestic labour force. For instance, in Britain – as elsewhere – waged domestic labour is gendered as an overwhelmingly female occupation. As such, it might appear appropriate to think about the potential pool of waged domestic labour in Britain as all women, and to account for the satisfying of demand in terms of general trends within women's employment. It could, for example, conceivably be argued that the increasing concentration of women in low-grade service sector jobs through the 1980s (Crompton and Sanderson, 1990; McDowell, 1989), the strong associations between this work and domestic work within the household, and the poor employment conditions found generally within such work (Crang and Martin, 1991), could make waged domestic labour appear relatively attractive to growing numbers of women. Alternatively, it could be argued that increasing levels of employment in waged domestic work reflect a decline in working-class female employment (Watson, 1991). However, we do not consider such broad brush generalisations to be valid. Our reasoning reflects the way in which our research has suggested *particular categories of waged domestic labour in contemporary Britain to be characterised by distinctive types of female labour.*

Unlike the situation in a number of advanced industrialised countries, waged domestic labour in contemporary Britain is not identifiable with a relatively homogeneous group of women. As we saw in Chapter Three, the contemporary waged domestic labour force in, for example, the United States, is composed predominantly of migrant women of colour. There, whilst certain groups of women – for example, Colen's West Indian day care workers or Glenn's generations of Japanese-Americans – have moved up the female occupational hierarchy, they have been replaced as waged domestic labour by further groups of migrant women of colour, for instance, by Puerto Ricans, Chicanas and Filipinas. In this case, the continued availability of a pool of waged domestic labour is linked to female migration. In comparison, in contemporary Britain, no such close association exists between ethnicity, female migration and waged domestic labour; although anecdotal evidence suggests that in London at least, if not in our study areas, certain households are using migrant women of colour as waged domestic labour.[1] Such differences can be accounted for in terms of the highly specific ways in which the labour of migrant women of colour has been used in Britain in the post-war period (Phizacklea, 1982, 1987). Thus, Afro-Caribbean women, much like Afro-Caribbean men, were used initially to 'stop' labour force shortages in the expanding public sector service industries of the 1950s and 1960s and in manufacturing assembly (Bryan, Dadzie and Scafe, 1985; Phizacklea, 1982, 1983; Stone, 1983). Likewise it was into definite labour market niches that Indian and Pakistani women were incorporated during the post-war period – principally into manufacturing industry, predominantly the textiles and clothing industries (Westwood, 1983). In terms of the resurgence in demand for waged domestic labour in Britain in the 1980s this has meant that no one clearly identifiable group of women has been available to satisfy the demands of middle-class households for domestic labour. The waged domestic labour force, therefore, is one which draws on a range of differences between women, notably those of class and the life-cycle. As we now show, these differences have combined in various ways to suggest an association between different forms of female labour and particular categories of waged domestic work.

Appendix 5.1 details the descriptive characteristics of the four sets of waged domestic workers whom we interviewed in our research. Although we distinguish between our north-east and south-east case studies, one of the obvious points to emerge is the minimal difference between our respective nannying and cleaning labour forces. Thus, in both the north-east and the south-east, the cleaning labour force was characterised in the main by older, usually married and solidly working-class women. In comparison, the nannies whom we interviewed were predominantly young – for the most part under 25 – and unmarried. Their social class background too was far more heterogeneous than that of our cleaners. So, whilst a minority of our respondents came from households in which both parents were in manual employment, most came from one of two cross-class households; namely households where the father was in low-grade management and the mother in some form of clerical work, and households where the father was in skilled manual work and the mother again in secretarial/clerical work. Such findings suggest that the nanny in contemporary Britain is an occupational category characterised predominantly by young, unmarried women from white collar, intermediate status households, whereas cleaning is the domain of older, married, working-class women.

Beyond this, it is important to stress two further points about the waged domestic labour forces whom we interviewed. Firstly, the vast majority of the nannies interviewed had a professional qualification relating either exclusively or partly to their work. This suggests that the nannying labour force in contemporary Britain is predominantly qualified. In comparison, cleaning carries no qualification tags and would appear to be an occupation in which literally any woman can be employed. Secondly, it is evident from Appendix 5 that our interviewed labour forces contained a high proportion of women who have been working, either as nannies or as cleaners, for a period of years and for whom this form of work represents their only paid employment. Given this, we suggest that, whilst working as either a cleaner or a nanny can be a temporary, casual and even – in the case of cleaning only – a supplementary form of paid work (ie., a second/third job), at the same time, such employment is something which other women do on a relatively permanent basis. In what follows we concentrate on the latter group, although we refer to more transitory workers where comparisons are illuminating.

Permanence raises the question of the relationship between demand for waged domestic labour and labour supply. In the following section we consider this relationship in as much detail as our research permits.

DEMAND AND SUPPLY IN LOCAL LABOUR MARKETS

As our analysis of demand for particular categories of waged domestic labour in Britain through the 1980s showed, there are both national and local labour markets for particular forms of waged domestic labour. We considered the first of these in Chapter Two, highlighting the importance of young migrant female labour from the regions in satisfying middle-class London households' demand for nannies. Beyond London, however, demand for all forms of waged domestic labour would seem to have been met primarily through local labour markets. Thus, in our interviews with employers in both the north-east and the south-east, we encountered only a handful of households – all in the south-east – which had recruited beyond the local labour market. Moreover, although restricted in a numerical sense, our interviews with employers suggested various points about the relationship between demand and supply within our study areas. Firstly, employers' comments on finding paid domestic help showed that although few encountered problems in filling nanny and cleaner vacancies, our study areas differed in terms of labour availability. Thus, in the north-east, households reported little difficulty in finding people willing to work either as nannies or as cleaners. Indeed, with the former category, choosing from a number of interviewees was the standard procedure. In comparison, for our south-east households things were more difficult, particularly in relation to cleaners.[2] Secondly, in discussing those whom they had employed and currently were employing, it became clear that employers were drawing on both newcomers to nannying and cleaning and those who had been doing this over a period of years. Such comments are important. For one, they support our contention that a degree of permanence has been a key characteristic of the nannying and cleaning labour forces in Britain in the 1980s. But, they also suggest that it may be accurate to think in terms of a numerical expansion in these two labour forces in the same period.

The characteristics of our respective cleaning and nannying labour forces have major implications for the analysis in the remainder of this chapter. Indeed, the existence of such clear differences between the two labour forces is taken to mean that it would be theoretically inappropriate either to treat these as a homogeneous entity or to think about a general waged domestic labour force. Instead, we need to think in terms of particular categories of waged domestic labour and to focus on how and why particular types of female labour have been so strongly associated with cleaning and nannying in Britain in the 1980s. Furthermore, our comments on labour supply suggest that we also need to consider the expansion issue. In what follows we examine the conditions and circumstances which have mediated the labour force participation of older, married working-class women and of young women with childcare qualifications in contemporary Britain. As we show, these conditions and circumstances are indicative as to why – in general – our employers encountered little difficulty in finding women willing to work as these particular forms of waged domestic labour. We begin our analysis by looking at the trends and circumstances which have had a crucial bearing on the labour force participation of older, married working-class women.

The older working-class woman and the labour market in the 1980s

Whilst derived from an extremely small number of interviews, the characteristics of our cleaning labour forces resonate strongly with what is known about women 'who did' in earlier decades of the post-war period in Britain. We can all, perhaps, recollect examples of older, married working-class women doing the odd spot of cleaning. And certainly both of us can recall instances from our own childhoods in the late 1960s and early 1970s of women from this social class who 'did' for other households, whether these were women from our own families or from those of neighbours. We do not therefore want to make any extravagant claims about the composition of the cleaning labour force in Britain in the 1980s. In our view this is probably much the same as it always has been. Such labour force characteristics are largely explicable in terms of the place of private domestic cleaning in the female occupational hierarchy. Given that the work itself is unskilled, physical and frequently dirty; that it is associated with other people's mess and consequently carries notions of personal servicing; and that it is work which is performed in an unwaged capacity by women, private domestic cleaning as an occupational category is located at the bottom of the female occupational hierarchy. As such, it is associated with precisely those women whose class, ethnicity, life-cycle stage and education combine to place them at a particular disadvantage *vis-à-vis* the labour market. For the most part it would be anticipated that such women would, be migrant women of colour. But, as we argued above, the highly specific nature of such women's incorporation within the labour force in post-war Britain has meant that it is class and the life-cycle, rather than ethnicity, which play critical roles in the production and reproduction of the private domestic cleaning labour force. For older, married working-class women in Britain, then, private domestic cleaning has traditionally offered one of a limited range of occupational possibilities.

Whilst we can consider the situation outlined above characteristic of the post-war period in Britain, a particular conjuncture of trends in Britain in the 1980s served to contract the range of labour market possibilities open to older working-class women and to reduce their labour force participation. This we feel to be of considerable significance in the creation of a potential pool of cleaning waged labour. The key trends which we isolate and discuss are: general trends in employment for older working-class women; the relationship between the older working-class household and economic restructuring; and the changing policy context of 'care'.

Employment, economic restructuring and older working-class women in the 1980s

As is well known, much of the early feminist research on women's employment concentrated exclusively on working-class women and their labour force participation within manufacturing industry (Cavendish, 1982; Coyle, 1982; Pollert, 1981; Wacjman, 1983). Whilst such emphases have since been corrected by studies of non-manual employment (Crompton and Jones, 1984; Downing, 1980; McNally, 1979), as well as supplanted by work which has emphasised the place of gender in the construction of specific occupations (see, for instance, Cockburn, 1983, 1985; Game and Pringle, 1983; Pringle, 1990; Walby, 1986; Witz, 1992), these early studies were, in a sense, right to focus on women's participation within manufacturing. At that time, particularly for women in the peripheral regions, manufacturing was the obvious employment choice. Apparently secure, characterised by a variety of shift patterns which could be combined in various ways with the demands of family and home, and relatively well paid, 'working on the line' offered the first real form of paid work beyond the home for many women in such places (Lewis, 1983; McDowell and Massey, 1984). Moreover, for other working-class women – notably those of the north-west and from working-class areas of London – manufacturing industry was the sphere in which historically women had been able to obtain employment (Savage, 1985). However, the restructuring of the 1980s changed all this.

It is, of course, impossible to summarise here the multiplicity of changes which occurred in the British economy in the 1980s. Instead we highlight two changes which have been important in shaping the labour force participation of older working-class women. Firstly, although it is the massive contraction in male employment within manufacturing which has received most academic attention, women also experienced manufacturing job losses through the 1980s (McDowell, 1989). In itself this is less noticeable than male job losses, in part because of the ease with which women can be switched to short-term and/or part-time working patterns (Beechey and Perkins, 1987; Rubery and Humphries, 1988). A more important reason for the relative invisibility of women's job losses in manufacturing, however, is the expansion in service sector employment through the 1980s. Thus, and as most overviews of women's employment in the 1980s stress, it was women, rather than men, who became increasingly concentrated in the burgeoning service industries of the 1980s. Such aggregate level changes suggest that, whilst

127

older working-class women might have lost jobs in manufacturing in the 1980s, they ought to have been able to find new forms of employment in the service industries. However, we consider this generalisation to be a considerable over-simplification of the reality for many such women, and for the following reasons.

In part the arguments which follow are speculative, requiring detailed case study research to substantiate fully. None the less, we consider recent research on women's employment, notably that of Rosemary Pringle on secretarial work (Pringle, 1990), to support some of our more assertive comments and arguments. Thus far in the literature on women's participation within the service industries, there has been seemingly universal acceptance of the assumption that it is women who, by virtue of their gender, are uniquely suited to employment within the service industries. The nature of service work itself – specifically servicing, but the similarities too which the work has with tasks which women perform in the home – is assumed to resonate strongly with women's traditional domestic role and their assumed subservience/ deference within patriarchally constructed gender relations. Whilst we agree in general with this assumption, we feel that specific service industries seem to be characterised by labour forces comprising different categories of women. One only has to think, for example, of the retail sales staff associated with certain clothes shops, or those used by large department stores to market and sell perfumeries and compare them with the staff of residential care homes or the schools meals service. In the first case it is young 'sexy' women who constitutes the labour force, and whose image plays a central role in the niche marketing of the particular commodities 'for sale'. In contrast, very different women, with a very different image, are associated with the latter group of service occupations. Here the older, mature 'mother figure', rather than the sexually alluring stereotype, is important, precisely because it is caring – as opposed to glitz and glamour – which characterises this form of service work. Given this, whilst working-class women in general may have moved from employment in manufacturing to employment in service industries in the 1980s, working-class women at different stages of the life-cycle will have been pushed towards very different forms of service sector work. For younger working-class women it would seem fair to suggest that it is retail sales, hotels and catering, fashion and beauty related services and the leisure industries which will have provided most of the employment possibilities in the 1980s. For older working-class women the employment choices would seem to have revolved primarily around commodified forms of domestic work, for example, privatised cleaning services; kitchen domestic work and care work in relation to the elderly and mentally and/or physically disabled.

The concentration of older working-class women in pseudo-domestic service sector work through the 1980s was something which we encountered in the life history data pertaining to our cleaning labour forces. Thus, whilst some of our cleaning respondents – particularly those at the top end of the age range – had been working as private cleaners for a period of years, others were relatively new to cleaning. Of this latter group, most had worked in manufacturing industry on leaving school, and had remained in this sector until the birth of their first children. Typically, they had then either withdrawn from the labour force for the child-rearing period, or worked intermittently, and in a part-time capacity. Most had then returned to paid employment either on a full-time basis or on a more substantial part-time basis than previously. However, it was then that they began to encounter redundancy in manufacturing. By the mid 1980s, almost all of our 'younger' group of cleaners had switched to a form of pseudo-domestic work. This we can see from the following two case studies.

Sheila[3]

At the time of the interview Sheila was 50, married, with three grown-up children. She left school at 15 with no formal qualifications and went to work full-time in a local dress factory as a machinist. By 19 she had become a supervisor; at 21 she married. At 23 (1963) she had the first of three children and left work. Whilst the children were very young Sheila did some part-time work in the same factory but it was not until the youngest reached school age (the mid 1970s) that she returned to full-time employment. Initially she went back to the company for whom she had worked previously, but then she got fed up with this and moved on to another machinist job. When this factory closed (approximately eighteen months later) she moved to another dress factory. In 1980 this closed down and Sheila, as a temporary 'stop-gap' measure, took a job as a part-time cleaner in a social club. She soon tired of this and took a kitchen domestic job in a local skills centre – a temporary move which lasted two years. In 1982 the centre closed, making Sheila redundant for a third time. She returned to work in another textile factory. Eighteen months later (1984) this too closed down and Sheila found employment as a night office cleaner.

Beryl

Beryl was 58 at the time of the interview and divorced. She has seven children. She left school at 14 with no educational qualifications and, at her mother's suggestion, went to work as a full-time domestic for a professional family. It was 1947. She stayed in this job for two years before taking another job as a machinist in a factory. However, she quickly moved on to work as a shop sales assistant, a job which she kept until marrying in 1952. For the next thirty-one years Beryl was occupied in a continual round of pregnancy, child-rearing and domestic labour. However, in 1983 when she was 49 she and her husband separated, making it essential for Beryl to return to full-time employment. The only work which she could find was as a full-time care assistant in a local residential home for the elderly.

The apparent concentration of older working-class women in pseudo-domestic service sector work in contemporary Britain is something which could be of considerable significance *vis-à-vis* the creation of a potential pool of private domestic cleaners in the 1980s. On the one hand, even by the standards of the service sector, such forms of working are widely recognised to be characterised by rates of pay amongst the lowest available to women; on the other, as a number of our respondents suggested, the social relations of many of these pseudo-domestic service industries underwent fundamental and radical revision in the post-privatisation climate of the mid-to-late 1980s (Coyle, 1985). From the perspective of the individual worker, it could be envisaged that, by comparison, private domestic cleaning – with its apparently personalised social relations, roughly comparable hourly rates of pay, and in many senses similar work demands – might appear relatively attractive. Middle-class households therefore, may have been able to satisfy their increasing levels of demand for cleaners through the 1980s precisely because, in the context of the 1980s, the work which they offered seemed more attractive than the alternatives within the formal labour market.

Whilst the above provides a powerful conjuncture of forces in the creation of a potential pool of private domestic cleaners in Britain in the 1980s, it is not the only influence. As important to acknowledge is the marital status of these older working-class women, and what has been happening to working-class households in relation to employment through the same period.

Some of the most influential social research to emerge in the 1980s is that which focused on households and economic restructuring (Harris, 1985; Morris, 1985a and b; Pahl, 1984). Such work has since spread in various directions (see, for example, the social polarisation and under-class debates (Bagguley and Mann, 1992; Morris, 1988; Morris and Irwin, 1992; Pahl, 1988; Robinson and Gregson, 1992). However, at the heart of the original work were questions to do with the changing relationship between households and work. As both Pahl's Sheppey-based research and Morris's studies in South Wales and Hartlepool showed, deindustrialisation in the peripheral regions, and in other areas reliant on male employment in manufacturing, brought with it fundamental changes in the relationship between households and waged work. Those households where the male partner remained in formal employment in the 1980s were shown to be increasingly likely to be characterised by dual, or even multiple, patterns of earning. In marked contrast were those households characterised by long-term male unemployment. Whereas male redundancy was shown to have no immediate effect on the labour force participation of female partners, long-term male unemployment, and the enforced household reliance on Social Security benefit, spelt a necessary withdrawal from the labour force by the female partner. Such households were shown to be increasingly tenuously related to the world of formal employment as the 1980s progressed. Indeed, as Morris's Hartlepool research demonstrated, for such households, short periods of casual 'off the cards' work were the most that could be expected to punctuate the giro economy (Morris, 1987).

For the most part, research on 'off the cards' work has emphasised the relationship between men and this pattern of working. Thus, in Morris's research, informal working by men is shown to fulfil two essentially social functions. Firstly, it is seen to provide a means of disassociation and differentiation from the domestic sphere and women. Secondly, it is seen to act as a source of meagre additional earnings which finance a degree of male social activity, notably drinking (Morris, 1987: 99). In comparison, whilst informal work for women is acknowledged to exist, Morris argues that – within her households at least – it was apparently less common (Morris, 1987: 100). This she relates to the conventional form of the gender division of labour in Hartlepool, and to established norms about gender roles and identities. Whilst the general validity of Morris's findings on women's participation levels within the informal sector remains to be confirmed, it may be premature to see women simply as minority participants in informal sector work. Rather, perhaps, it is the differences between male and female participation within the informal sector which may be important. Thus, whereas for men informal work is located beyond the domestic sphere, for women, the latter has long functioned as an important source of casual earnings. Indeed, activities such as the making and icing of wedding, birthday and Christmas cakes, knitting and sewing, as well as the more basic labour intensive tasks such as cleaning and ironing, have long been supplementary earners for working-class women.

The above research has shown that although a growing number of working-class households were becoming cut off from the world of formal employment in the course of the 1980s, it was households beyond the immediate child-rearing life-cycle stage who were most affected by deindustrialisation. As manufacturing industry rationalised operations through the 1980s, it was men in the 40-plus age bracket who proved

131

most vulnerable to redundancy. We can anticipate then that it was middle-aged working-class households which were most affected by the manufacturing 'shake out' of the early 1980s, and which have been particularly disadvantaged in the radically altered labour market of the late 1980s and early 1990s. In such circumstances, it is not difficult to envisage the emergence of household work strategies involving inter-mittent casual working by both male and female partners. Private domestic cleaning represents an obvious source of informal work for such female partners.

Working class women and the changing policy context of care in the 1980s

> We know the immense sacrifices which people will make for the care of their own near and dear – for elderly relatives, disabled children and so on. . . . Once you give people the idea that all this can be done by the state . . . then you will begin to deprive human beings of one of the essential ingredients of humanity – personal moral responsibility.
>
> (Margaret Thatcher, 1978, quoted in Croft 1986)

The 1980s, much like the previous decade, saw the continued espousal in social welfare and related policy circles of the concept of 'community care'. As Finch and Groves (1980) indicate, the term 'community care' was initially used in the 1960s to refer to a shift in the location of care, from remote institutions to small-scale residential units situated in urban and suburban areas. However, by the mid 1970s, care in the community had become equated with care by the community. Increasingly, in a context of contracting public sector expenditure, the potential for expanding social service provision through informal care had become realised. The advent of the New Right and Thatcherism, however, added further twists to what had essentially been a policy legitimated by debate on public sector expenditure levels. Debate on public sector spending became overlaid and recast by the ideologies of individualism and anti-collectivism; by discourse on the primacy of the nuclear family as the basic social unit; and by a re-assertion of individual moral responsibility (Dalley, 1983). Thus, it was not just cost-based arguments which legitimated the increased emphasis on care by the community, but the entire ideology of the New Right. The discourse of the 1980s saw the assertion of a fundamental dualism: on the one hand was collective care (costly, poor quality and morally irresponsible); on the other, that which could be provided by the family (cheap, good quality and morally superior). And, as Dalley shows, these changes were experienced by a number of groups, for example, the elderly, mentally handicapped and children.

132

As much feminist work on care and caring has emphasised, the term 'community care' is essentially a euphemism for care by women (Finch and Groves, 1983; Graham, 1983; Ungerson, 1983). Some of this work, for example that of Graham and of Ungerson, has explored more fully, and in an explanatory sense, the gendering of care, but more central to debate have been considerations of alternative feminist forms of care (Finch, 1984; Finch and Land, 1981); policies for carers (Croft, 1986); and accounts which stress the interdependence between community care and women's participation within the labour force (Finch and Groves, 1980; Finch and Mason, 1990; Land, 1978). It is the latter strand of debate which is particularly pertinent to our concerns. As long ago as 1980, Finch and Groves were spelling out the implications of community care for women's labour force participation. They remark, for example:

> The present combination of expanding community care and declining employment could well result in strong forces pulling women back into domestic, caring roles. . . . [F]rom the viewpoint of the individual woman, the situation will be that she sees no real alternative to providing care herself: it is an obligation laid on her by the cultural definition of women as carers and by the lack of other provision. . . . This raises the very real possibility that women may be expected to and may feel that they have to give up work in order to provide continuous care for a dependent relative.
>
> (Finch and Groves, 1980: 503–6)

More recently, in the course of a study of the care of elderly dependents, Finch and Mason (1990: 362–3) have argued that the assumption that the onset of caring responsibilities necessitated an automatic withdrawal from the labour force by women was overly simplistic. Instead they argue that increasingly women in Britain will construct compromise positions which enable them to satisfy their obligations regarding adult kin and to remain in some form of employment.

The 1980s, then, saw the continuation of community care policies and increasing ideological pressure on women to fulfil the caring needs of dependants and relatives. Not only was it that, increasingly, no alternative form of care was available to satisfy such needs, but morally too 'there was no alternative'. Engaging centrally and emotively with women's identities, such policies were able to compel women to feel obliged to care if the situation demanded. However, whilst it is clear that these policies affected women in general, they affected specific groups of women in particular. There is now unequivocal evidence to show that the British population is an ageing one, and that it is the elderly who are placing the greatest numerical burden on health and social services. In a context where 'family care is best', and in which major cuts have also occurred in support and ancillary services – for example, meals on wheels and home helps – it is women currently in their forties and fifties on whom the burden of care for the elderly is falling. A second expanding area of home-based care is that associated with the care of sick and/or disabled partners. In this context we would suggest that it is social class which is critical in determining individual responses to caring needs. For women working in service class occupations there are powerful reasons for remaining in full-time employment. On the one hand, the financial resources of such households are sufficient to purchase replacement care (Glendinning, 1991); on the other, whilst the obligations to care are still felt, counter pressures from the place of work within such individual women's identities are strong. In contrast, for those women with patchy, intermittent labour market histories, whose primary identity has revolved around being the wife/mother figure, there frequently is no alternative but to withdraw from the labour force. Indeed, the labour market position of such women confines them to low-paid forms of work, none of which have the earning capacity to support private sector replacement care.

As Glendinning's work has shown, the financial implications for those households in which women withdraw completely from the labour market to become full-time carers are considerable. Albeit for different precipitating reasons, these households, through their enforced dependency on Social Security benefit, find themselves in the same difficult financial circumstances as those comprising unemployed partnerships. Much as with the latter, we can envisage that private domestic cleaning offers a potential coping strategy for at least some of those households where the female partner withdraws from the formal labour market, and for two reasons. Firstly, not all of those being cared for will require full-time twenty-four-hours-a-day care. Though their care needs may be sufficient to necessitate a withdrawal from the labour market, it may still be possible for the carer to leave the house for periods at a time. Secondly, whilst long-term dependency on Social Security benefit spells living on the poverty line, it also restricts formal labour force participation. For households, therefore, in which the full-time female carer is able to leave her dependant(s), it is the informal sector which offers the main opportunities for supplementing household incomes. As we have already argued, such opportunities for women stem primarily from the domestic sphere. Private domestic cleaning, then, along with a range of other domestic tasks and/or activities, can be seen to offer one of the few potential ways in which women carers can counter the enforced poverty which characterises their lives in contemporary Britain.

The foregoing establishes the major trends and circumstances which have conjoined in the 1980s to shape the labour force participation of older married working-class women in Britain. The concentration of employment opportunities in commodified forms of pseudo-domestic work; continued high levels of long-term male unemployment; and the ideologically and economically legitimated policy of community care have combined either to reduce the range of formal labour market possibilities available to this particular group of women, or to prohibit formal employment altogether. Moreover, two of these trends have condemned such women to benefit dependency and its consequence, living on or below the poverty line. In an abstract sense we would suggest then that in such contexts private domestic cleaning makes sense. Compared to pseudo-domestic work in the post-privatisation climate, it could well appear attractive. But it is the second category of trends which we feel to be potentially of greater causal significance. Informal, invisible and characterised by highly flexible working hours, private domestic cleaning offers one of the few ways in which working-class women can supplement household incomes dependent on Social Security. Benefit-dependent households, whose numbers increased considerably through the 1980s (Field, 1989; Mack and Lansey, 1985; Pond, 1989), and the women within them, would seem to constitute a clear source of private domestic cleaning labour for middle-class households in contemporary Britain. In the following section we examine whether these potential tendencies are realised within our cleaning case study labour forces, and if so, why.

PRIVATE DOMESTIC CLEANING AND BENEFIT DEPENDENCE IN CONTEMPORARY BRITAIN

The most striking point to establish at the outset about our case studies of private domestic cleaners is that the majority were from households which, for one reason or another, were dependent on Social Security incomes. Three categories of household proved to be particularly common: those with a male partner in receipt of invalidity benefit; those with an unemployed male partner in receipt of income support; and those reliant on the state pension. The following case studies provide examples of all three instances.

Households on invalidity

Doris (private domestic cleaner for two years at interview)

Doris is 46, and married with three grown-up children. She has had a succession of pseudo-domestic jobs in her working life, some of which predated her marriage. Most, however, punctuated the child-rearing years. In 1984 she moved from part-time employment as a kitchen domestic in a local hospital to a part-time job as a ward domestic. In

135

1985 her husband – then in skilled manual work – began to encounter serious health problems and subsequently was made redundant. Doris gave up her job to care full-time for her husband. Although, in time, Doris's husband made a satisfactory recovery, he is medically unfit for work. This situation posed major financial problems for the household. As Doris expressed it:

> It was like a bottomless pit where money was concerned. So I had to get out and make some somewhere. I mean, I'd spent a lot of money on the garden . . . the little bit of reserves we had without touching the solid bit had gone. So I thought, 'I'm going to have to do something here.'

It was then 1988. The household had by then been dependent on Social Security payments for two and a half years. Doris's husband was unable to do any form of work, so the household was reliant upon her to supplement their income. Doris's response was private domestic cleaning, initially for more than one household.

Sheila (private domestic cleaner for three years at interview)

In 1984, Sheila was working as a night office cleaner. In 1986 her husband – then in manual work – had a major coronary and Sheila felt compelled to leave her part-time employment which, by her own admission, she loved, in order to become a full-time carer. As she says:

> I had to leave because there was no way I could leave him in the house. Like I was going out at four thirty and coming in at nine at night. And he didn't like it. If anything happened to him, you know.

With time, Sheila's husband gained in confidence and she felt able to leave him. However, much as with Doris's household, it was the financial difficulties of benefit dependency which, by 1987, had started to bite:

> And I thought, 'What shall I do?' I've always been used to me own money . . . every time I've worked, what I've earned, I've always had me own money. He never said, 'That's my money.' When the kids were little, fair enough, it went to buy shoes and the like. But when they were older and started working, me pay was mine to do what I wanted. So I says, 'I think I'll try for a job like.' So he says, 'Well it's up to you Sheila.' 'Cos we were both sitting in the house looking at each other!

Much as with the previous case, for this household private domestic cleaning provides a means of supplementing benefit. However, it has also provided Sheila with financial independence from her husband –

something which has been important to her throughout her working life.

Households on income support

Iris (private domestic cleaner for over fifteen years at interview)

Iris is 52, married and has two grown-up children. In many respects Iris's employment history is representative of the classic working-class experience. She left school at 15 and went to work full-time in a curtain factory. Married at 19, she continued to work in this factory for a further year before leaving to work in another factory assembling radios. This lasted for a further eighteen months. She then became pregnant (1959) and took ten years out of the labour force for childrearing. When her youngest child was 9 (1971), she returned to part-time employment as a home help in the local area, a job which she acquired through her mother who was, at the time, a home help. Iris worked as a home help for thirteen years, gradually mixing her local authority people with a number of 'privates'. As her children grew older, she increased her hours in paid work, so by the early 1980s she was working both as a home help and for three private employers. Then, in 1982, two factors intervened to change this situation completely. Her husband, an un-skilled manual labourer, was made redundant and her mother became mentally ill. As a result Iris gave up her employment as a home help. She also gave up all except one of her private cleaning jobs. Iris's mother died two years later but her husband is still unemployed. The coping strategy which this partnership has developed is one where both partners do 'off the cards' work. For Iris's husband, such work is intermittent. For Iris, private domestic cleaning is a regular means of supplementing income support, as well as something which she has done for years.

Joyce (private domestic cleaner for over eight years at interview)

Joyce is 50, married and has six children. She left school when she was 15 and for the next two years worked full-time, firstly as a machinist in a dress factory and then as a sales assistant in various shops. At 17 she married and, like Iris, spent ten years out of the labour force in a round of sequential pregnancy and child-rearing. She returned to part-time employment in 1968. Over the course of the next ten years she worked in various pseudo-domestic jobs – either as a cook or as a cleaner. In 1979 her husband – a skilled manual worker – was made redundant. He has been formally unemployed ever since. By 1984 the enforced house-hold reliance on benefit was causing considerable financial difficulties.

137

As with Iris's household, this partnership's financial coping strategy is for both Joyce and her husband to take 'off the cards' work. For Joyce, like Iris, private domestic cleaning offers a regular means of supplementing income support.

Households in receipt of the state minimum pension

Edie (private domestic cleaner for four years at interview)

Edie is 63 and married. She has one son. Her husband is retired and the couple are in receipt of the state pension but have no other source of household income beyond a small amount of life savings. Private domestic cleaning offers one of the few strategies open to them for 'helping out' with the pension. However, what is particularly interesting about this household is the way in which Edie moved into private cleaning:

INTERVIEWER: So why did you choose to do cleaning then?
E: Well I just happened to be talking to 'Celia' on the street and I happened to say that I'd packed me job in (she was then 59). And she said, did I fancy doing cleaning – because I just lived next door to her then. So I said, 'Oh I don't know.' And then I thought, 'Well it'll get me out. It would give me something to do.' And now the extra money comes in handy. So that's how I got into it and I've been here ever since.

Edie became a private domestic cleaner once she left a part-time job in the formal sector, but only because she was approached by a near neighbour. Now in different financial circumstances, the *raison d'être* for cleaning has changed. What was once an occupation which got her out of the house has now become a matter of financial necessity.

Rose (private domestic cleaner for over eight years at interview)

Rose is 72 and widowed. Her husband died in 1977. Rose left school at 14 and went for six months' training as a domestic. She then took employment as a domestic in a school. It was 1933. She stayed in this job until her marriage at 22 in 1940. Her daughter was born in 1945. Once her daughter was 5 Rose returned to part-time work in various forms of domestic employment. After a sequence of such jobs lasting over thirty years, Rose was made redundant in 1982. She was then 64 and reliant financially on a combination of her state pension and a war pension. It was this set of financial circumstances, and her wish to be in a position to provide 'extras' for her grandchildren, which led her to seek work as a private domestic cleaner. The latter is a form of the only type of work which Rose has ever known and provides her with the financial resources both to ease her own financial dependency on the state, and to fulfil the role of grandmother.

The picture which emerges here is remarkably consistent. From the first pair of cases we can see how private domestic cleaning connects with those working-class households dependent on invalidity benefit. In both of these cases, the 1980s witnessed the onset of major caring responsibilities for the female partner. Whilst neither of the male partners in these households required full-time care, their needs in the policy context of community care were – at least initially – sufficiently great to require their partners to withdraw from the labour force. Interestingly we can see, particularly in the case of Sheila, how – and in spite of her obvious enjoyment of paid work – she felt obliged to become the full-time carer. Of even greater importance though is that for these women – unlike Finch and Mason's respondents with elderly dependants – household reliance on invalidity benefit meant that there was no option but to withdraw from the formal labour market. The result in the longer term was increasing financial hardship, which only the female partners in these households were in a position to do anything about. Private domestic cleaning for these households represents a specifically gendered response to such circumstances. Not only does it offer a means of supplementing meagre benefit-dependent incomes; it also alleviates the financial dependence of the female partner on the benefit received by their male partners, and is a means of release from the pressures of being the carer. This situation is one which resonates strongly with that of households with unemployed male partners but with subtle, important differences. Once again, private domestic cleaning represents a gendered response to long-term household dependency on a form of Social Security benefit. But, importantly, it is not overlaid by the social context of caring. A further difference is suggested by the existence in our respondent unemployed households of clear partner strategies to supplement benefit-dependent incomes. Undeniably, private domestic cleaning in the above two households had offered a regular and long-term source of 'off the cards' work for the female partners. In comparison, the 'off the cards' work available to their partners was more casual. The one therefore provides a relatively secure source of household finance over and above benefit; the other is a much more unpredictable source. Moreover, in the light of Morris's findings on the different uses to which male and female earnings from informal work are put (Morris, 1987: 101), we can anticipate that the income earned through private domestic cleaning by these women not only augments benefit but plays a critical part in household budgeting.

139

Finally, there is the slightly different case of private domestic cleaning by female partners within households dependent on the state pension. As with the previous two household categories, the major reason for private domestic cleaning is financial. Such women are, at root, supplementing household incomes which they deem to be inadequate. However, in so doing they are also providing the financial resources for necessary 'extra' household expenditure.

We consider the above material to be extremely significant. It appears that the decision of older, married working-class women to clean within households has very little connection with comparisons made with alternative forms of paid employment. Instead, for the majority of those cleaners whom we interviewed, the decision was bound up with benefit dependence of one form or another. Our interviews with women currently working as private domestic cleaners in both the north-east and the south-east suggest that at least part of the pool of female labour on which middle-class households drew in order to satisfy their demand for cleaners through the 1980s was composed of women from benefit-dependent households. However, we do not wish to claim that such households provided the only source of such labour. In the course of our interviews with employment agencies providing a private domestic cleaning service, it was suggested to us that a second category of female labour – also working-class, but at a different stage in the life-cycle – is to be found working in this capacity. These women are younger, married, and with husbands in employment. But their children are either of pre-school or primary school age. We did not encounter many such women in the course of our research but Sharon represents one such instance.

Sharon (private domestic cleaner for five years at interview)

Sharon is 36 and married with three children, two of whom are of school age. Her husband is a milkman. Sharon left school at 15 with no educational qualifications and went to work as a shop sales assistant. At 20 (1976) she married her husband, then a butcher. When her first child was born (1978), she withdrew from the labour force; shortly afterwards, the family took over a newsagents and Sharon helped out serving in the shop. In 1985 Sharon and her husband gave up the business and her husband then started 'on the milk'. At this time their second child was just 5. Sharon then found a job at the local primary school working in the canteen. Shortly afterwards she was approached by a teacher to work as a private domestic cleaner. She has been working for this household for five years. In 1987, with the birth of her third child, Sharon left her job at the school and currently takes her pre-school-age child with her when she goes to clean. Like Edie, Sharon got into private domestic cleaning because she was approached by someone. She continues to work in this capacity because she finds that it fits in with the kids; that it allows her to spend her evenings with her husband; and because she does not have to declare these earnings. For her, private domestic cleaning represents a means of supplementing household earnings whilst continuing to be a full-time mother.

As we emphasised initially in this section, private domestic cleaning has long been one of the main ways in which unwaged working-class housewives have supplemented household earnings. Sharon represents a clear instance of a woman doing this who is at one end of the age spectrum. At the other end, and unlike Edie and Rose, are women who have been working as private domestic cleaners for decades rather than years. The most extreme example of this encountered in our interviews was that of Win.

Win (private domestic cleaner for thirty-one years at interview)

Win is 71 and widowed. She was born in 1920 and married at 19 in 1939. She has five children. Win left school at 14 with no qualifications. She went to work first as a domestic and then in a biscuit factory. She stayed in this factory until her marriage to a fellow employee. Win did not work whilst her children were young, except for a few short stints on the night shift at the biscuit factory. However, in 1960 she answered an advertisement for a cleaner. She has been with this household ever since (i.e., for thirty-one years). Whilst her husband was employed, Win's work provided the household with a small, but useful, supplementary income. Now, as a widow and dependent solely on the state pension, Win – like Rose – finds the money useful to support 'extras'. In her case it goes on holidays and buying clothes.

The presence of women like Win in our interview programme, as well as that of younger women who only began private domestic cleaning in the 1980s, is of considerable importance. Indeed, it suggests a change in the composition of the private domestic cleaning labour force in Britain in the 1980s and a numerical expansion in labour supply. From what little is known of private domestic cleaners in Britain in previous decades, and through the oral histories of our own families, it would seem that it is the Wins of this world who have traditionally constituted the primary source of cleaning labour for middle-class households in Britain. Indeed, in many ways, in going out to clean in 1960, Win was doing exactly the same as Sharon in 1985, namely finding a form of work which topped up her household's income without interfering with the demands of being a full-time wife and mother. In one sense, albeit that the incomes which they are supplementing are derived from the

141

state rather than from waged work, the women from benefit-dependent households who began to work as private domestic cleaners in the 1980s are doing the same 'top up' type of activity. But in another sense these women are in a very different position. For those women whose households are reliant on either invalidity benefit or income support, formal labour force participation is severely restricted. Private domestic cleaning for them represents not a choice mediated by traditional gender roles and the conventional form of the gender division of labour, but one forced on them by the structure of the benefit system. Without a doubt, these are women who would be participating in the formal labour force if circumstances were different. Given this, we can suggest that the private domestic cleaning labour force in contemporary Britain is composed of two different types of women. On the one hand there are those like Win and Sharon; women for whom working as a private domestic cleaner either originally fitted in with being a full-time wife and mother or currently does so. On the other hand, there are those women like Sheila, Doris, Joyce and Iris, for whom private domestic cleaning is either a manifestation of the massive contraction in male employment within British manufacturing industry through the 1980s, or a reflection of the financially dependent status of female carers on their own dependants.

Given the increasing number of long-term unemployed households and female carers in Britain through the 1980s, we suggest that the private domestic cleaning labour force in Britain in the same period became characterised by increasing numbers of benefit-dependent women and by increasing numbers generally. With estimates of the number of individuals reliant on income support running at over 8 million in the late 1980s (Field, 1989) and related high levels of long-term male unemployment in the peripheral regions, it is not difficult to see why, in general, the middle-class households whom we interviewed experienced few difficulties in filling their cleaning vacancies, as well as why our south-east households encountered greater problems than our north-east respondents in finding numbers of women willing to work in this capacity.

Having examined the conjuncture of trends and circumstances which generated an expanding pool of private domestic cleaners in Britain in the 1980s, and having indicated how these appear to have modified certain of the characteristics of the private domestic cleaning labour force in post-war Britain, we now move on to examine the very different case of our nannying labour force.

FOR LOVE NOT MONEY: 'LEARNING TO BE MOTHERS' IN THE 1980s

As suggested by our interviews with employers, the popular image of the nanny is one of a trained woman in a uniform, in short, of a

'Norlander'.[4] It also suggests someone who lives with a family; even of someone who lives with a family for a period of decades. This image is at some considerable remove from the nannies employed by the new middle classes in the 1980s and early 1990s. For the most part – at least outside London – these nannies are day nannies. They are also far more likely to be the products of local technical colleges than of the private nanny training colleges of Norland, Chiltern and Princess Christian. Moreover, and as we show in Part Two, whilst the attachment of these young women to middle-class families and their children can be considerable and long-lasting, they are unlikely to find themselves in the pseudo-retainer position described by Gathorne-Hardy (1972). In what follows we examine how and why the young women whom we interviewed, all of whom at the time of interview were working for service class households, came to enter the labour force as nannies.

As will become clear in Part Two, the social relations of the nannying form of waged domestic labour in Britain in the 1980s, and the nature of the work itself, are structured by traditional ideas about mothering and the gendering of childcare. This has two implications for the composition of the nanny labour force in the 1980s. Firstly, it means that this labour force will be almost exclusively female. Secondly, it suggests that the occupation will be characterised predominantly by women whose gender identities are primarily traditional and conformist. The latter was apparent from our interviews with nannies and from our observational fieldwork. To work on a permanent basis – as opposed to on a temporary and/or casual one – with young, pre-school-age children, for long hours, for little material reward and over a period of years, is undeniably a 'labour of love'. It requires that the women who work in this capacity enjoy the activities and tasks associated with childcare; that they identify with, and enjoy the constant company of, young children; and that they feel positive about a form of work which identifies them so strongly with young children. The nannying form of waged domestic labour therefore is likely to be characterised by particular categories of female labour. Given the organisational structure of the occupation (see Part Two), it is possible to identify three potential categories. The first is a group of young women who are not mothers, but for whom children and their care figure centrally in their emergent identities. The second group comprises older women – themselves mothers – but whose family no longer makes major demands on them. A third group is a small subset of older women; those who identify strongly with children, but who are not mothers, that is the stereotypical image of the upper-class nanny in Victorian Britain (Gathorne-Hardy, 1972). In our research the first potential category was dominant. Indeed, we encountered only two of the second category in our north-east fieldwork and the same number in the south-east. Moreover, neither of us found anything remotely approximating to either a Nanny Everett or a Mary Poppins figure! Without doubt, such findings reflect our focus on the new middle classes in Britain in the 1980s. If we had looked at demand from other sources – for example, the aristocracy and the upper classes – we might have found something different. However, the expanding demand for nannies by the new middle classes through the 1980s seems to have been met in the main by the first potential category of female labour. As a consequence, any explanation of the production of the nanny labour force in Britain in the 1980s has to focus primarily on this group.

In what follows we argue that the production of an expanding labour force of nannies in Britain through the 1980s has to be understood, at least partially, in terms of the longer-term socialisation of a particular group of young women and their emergent gender identities. The argument which we develop has three steps. Firstly, we show how the emergent adult gender identities of our future nanny labour force were formulated through, and shaped by, sets of social practices occurring in three key sites, namely the parental home, the neighbourhood and schools. Secondly, we argue that one of the consequences of these practices – either intentionally or unintentionally – has been the production of identities which were, and are, conformist and traditional. Of course, neither these practices nor the identities which they shaped can be considered unique to the 1980s. However, such identities prove central to understanding the immediate choices which a specific group of young women made at 16-plus in Britain in the 1980s. Indeed, they are shown to have been critical in enabling the young women we interviewed to have rejected the other forms of labour force participation open to them in contemporary Britain, in favour of a vocational training course, either exclusively or partially concerned with childcare. Finally, we show how, once having been on such a course, the reality of contemporary childcare provision in Britain propelled these young women into seeking employment as nannies. We begin with a set of portraits which illustrate the emergent identities which are important here.

Typical girls, typical nannies . . . Rebecca and Janice . . . and Marie

Rebecca (nannying for nearly three years at interview)

Rebecca is 20 and lives with her parents and her younger brother, aged 13. Her father is a credit control manager and her mother works voluntarily at a local playgroup. Although not engaged, Rebecca has a steady boyfriend. Rebecca left school at 16 with a handful of Certificate of Secondary Education passes (CSEs). Before leaving school she applied for the two-year National Nursery Examination Board (NNEB) course at the local technical college but was rejected.[5] On the advice of a teacher, she took a place on a two-year course leading to the Preliminary Certificate of Social Care (PCSC) qualification. This was very much a 'try it and see' strategy: at this stage Rebecca had absolutely no idea what she wanted to do in the way of paid employment. Whilst she was on this course, Rebecca went to evening classes and gained Ordinary level and two Advanced level passes because she thought that she might like to be an infant teacher. On completing this course, Rebecca found a job as a nanny. She has been in this job nearly three years, and now cares for a 3-year-old and a young baby. During this period of employment Rebecca has applied twice for places on a B.Ed. course. Each time she has been accepted, but each time she has turned the place down at the last minute. Although she realises that the B.Ed. offers her the chance of a career, and that nannying is a 'dead-end' job, she cannot bring herself to make the break. Ultimately, when it came to making the final B.Ed. acceptance, she preferred the short-term security and attractions of working for her current employers – in her eyes, the regular wage and a new baby to look after. These decisions have to be understood in terms of Rebecca's aspirations. By 25 she hopes to be married; to have her own home and to be at home looking after her own baby. She wants, in other words, to be a full-time housewife and a full-time mother. Working as a nanny fits comfortably with this. It fills in the time and it does so in a way which actively complements her aspirations. In comparison, going away to college spells uncertainty, insecurity and a potential career which could sit uneasily with her desire to be a traditional wife and mother.

Janice (nannying for nearly three years at interview)

In many ways Janice is similar to Rebecca. She is 21 and lives with her parents. Her father is a heating engineer, and her mother works as a sales assistant in a major retail store. Janice is engaged. Her fiancé is a plant operator. Janice left school at 16 with CSEs and, like Rebecca, did the two-year PCSC qualification. She would have preferred to have done the NNEB course but did not have the qualifications for this. Since qualifying, Janice has had two nanny jobs. The first she left after eighteen months because of poor conditions. She has been in her current job for nearly eighteen months and looks after a four-year-old and a young baby. Like Rebecca, Janice plans to stay in this job for the foreseeable future, at least until she and her fiancé have sorted out when they are going to get married and the purchase of a house. She envisages that by 25 she will no longer be nannying and that she will be at home looking after her own children. She is, in her own words, 'a home bird' with entirely traditional aspirations.

Both Rebecca and Janice are typical nannies. Marie, in contrast, represents a different category of young woman; one whose passage through nannying is much more temporary.

Marie (nannying for eighteen months at interview)

Marie is 19 and lives with her parents and younger brothers, aged 17 and 6. Her father is a draughtsman and her mother a secretary. Marie is neither engaged nor does she have a regular boyfriend. Whilst she enjoyed being around young children as a teenager, Marie had not envisaged working with them. She had anticipated staying on at school to do A levels. However, her results did not match up to expectations, so she took a fall-back place on an NNEB course. She was one of the few respondents whom we interviewed who did not complete the course. Marie's reasons for this are interesting. For a start, Marie had been told at school, both by teachers and careers advisers, that she was 'too good' for an NNEB course. When interviewed in our research, she came across as a bright, intelligent and critical young woman. We did not, therefore, feel that such reported comments were at all inaccurate. Moreover, such assessments were backed up by Marie's own evaluation of the NNEB course. Although she enjoyed elements of the course – for example, book reviews and compiling children's poetry anthologies (i.e., the more academic content) – she assessed the course placements extremely critically. Indeed, she was one of the few young women whom we interviewed who had worked out the place of the NNEB qualification in the labour market while she was still on the course. We will come back to this point later. However, from the point of Marie's immediate future, such critical assessment led her to question the point of continuing on the course. In her own words:

> I'd figured out that I wanted to do nannying when I finished and I thought, 'Well I can do it quite easily now, so why not do it now?' Why waste another year and a half doing that when I can do it now?

Whilst on a nanny placement, the chance of a job came up and she took it. At the time of interview she had been in this job for eighteen months. But Marie had no wish to remain a nanny and, shortly after our contact with her, she left to go back to college full-time. This move reflected her assessment of nannying and of nannies. Nannying she labelled as 'playing at mother all day'; being a nanny, a job for women with limited goals and aspirations:

> I thought that this new generation were all going to be ambitious women wanting to be career women and things like that. But all I

146

seem to meet nannying are women who are just prepared to – well basically meet somebody and get married and have children and stay at home. When I meet them I can't believe it but they are. I know some people – I don't know – I suppose it's what you're born with – I mean some people haven't got, you know, the intelligence, the ambition to go far. And I think, well nannying is a great job for them. But I mean, I couldn't do it for that long because it's not sort of demanding enough. It's the same day in and day out – basically the same. And it's not that I've got anything against young children, but basically I think I'm too intelligent. I think I can do better.

Marie's comments pin-point the dominant characteristics of the nanny labour force and highlight the importance of marriage and motherhood to the identities of the young women who, like Rebecca and Janice, constitute the bulk of this labour force. They also show why her own employment as a nanny was relatively short-lived. Marie had had enough of 'playing mother'. She found the job insufficiently demanding intellectually, and she wanted to 'get on'. In the course of our fieldwork we met other young women like Marie. All were nannying as temporary 'stop-gaps', and mostly because, like Marie, they had failed to get the necessary grades first time round. Like Marie, they differentiated themselves strongly from the typical nanny, whom they dismissed as sitting around waiting to be a full-time wife and mother.

Our fieldwork findings and interviews with nannies and with employers corroborated entirely the picture painted by Marie. The nanny labour force in contemporary Britain is characterised in the main by young women with very traditional identities. Moreover, such identities are of critical importance in enabling them to do this form of work. Given this, if we are to account for how the young women whom we interviewed came to enter the labour force as nannies in the 1980s, we have to focus on the means by which these emergent identities were produced. In what follows we emphasise the formative role played by social practices associated with three key sites: the parental home, the neighbourhood and schools.

The parental home and the neighbourhood

The most striking initial points about our two nanny labour forces are that all came from households in which the mother had not worked in paid employment through the early child-rearing period, and that a significant minority had siblings considerably younger than themselves. We consider both points to be important. Taking the maternal pattern of employment first. In following the classic 'M' shaped pattern of female labour force participation in post-war Britain, the mothers of our future nannies were by no means exceptional (Dex, 1988; Martin and Roberts, 1984). However, it is what lies behind this pattern which is of particular interest to us. Indeed, when we questioned nannies on their own mothers' ideas about motherhood and employment, what was conveyed to us was a strong sense of 'a mother's place is in the home with her children'. This we can see in the following extracts. In the first, Kay outlines the highly traditional views of her mother. In the second, Clare articulates the ideas which lie behind her mother's current combination of paid employment and mothering.

Kay (23)

K: She wanted to be there for us.
INTERVIEWER: That was important to her?
K: Oh yes. She definitely, she still – and did then – felt that a mother should give up everything for the family. She gave up everything for me dad too you know.
INTERVIEWER: Really.
K: Oh yes. Definitely. She felt that she should be at home to be making meals, whatever. He's never helped around the house. That's a different thing I suppose, but she's always believed that a mother's place is with the children. Definitely. She really feels that it makes a difference to a child's life you know.

Clare (20)

Now remarried, Clare's mother has a seven-year-old child.

> She's got to work for the money. But she thinks it's right to fit it in with the children. She works Sunday from 1 to 10.30 p.m. That's a double shift so she gets paid double for that. And then Monday, 5 till 10.30. Well me dad's in for 5 so that works out brilliant you see. And then the rest of the week she's off. So she's got all day with Damon, to do the housework, everything.

In the first case what we see stated is the traditional female gender role and the traditional female identity: a mother's place is at home with her children. The second case represents an instance of the primary role (that of mother) shaping the form of labour force participation (part-time and in accordance with the work pattern of the male partner). There are, of course, temporal differences here. The situation which Kay describes is something which her mother felt in the late 1960s and early 1970s, although Kay suggests that these are ideas which her mother still holds. For Clare's mother, the late 1980s and early 1990s provide the historical context. Clare, however, is of a similar age to Kay, and when she was young her mother worked informally from home:

> She used to sew wedding dresses and bridesmaid's dresses for large people would you believe, who used to come in and wanted their clothes fitted for them 'cos they couldn't get the right sizes. And she did that while we were little. She worked from home for years and years – till we were old enough to watch ourselves.

Her views of combining mothering and employment therefore have remained fairly constant, much as Kay suggests of her mother.

It is difficult to gauge precisely the effect of these ideas and actions, both specifically on the individuals whom we interviewed, and more generally within our future nanny labour force. Indeed, in that such evaluations are predominantly psychological in nature, they lie beyond our concerns and our expertise. However, it is difficult to see how they can have been anything but formative. Firstly, these young women have all been socialised from a very early age to see women's primary role as that of wife and mother, and to see as the norm the identification of childcare with mothering. Secondly, their own mothers have promoted (through their own actions, through their explicitly articulated ideas, or a combination of both) the dominant ideology of childcare. Though it is entirely possible that such circumstances could produce rejections of such ideas, roles and identities, it is evident that in many cases they are, to a greater or lesser extent, accepted. Rebecca and Janice, for example, both envisaged leaving the labour force in order to look after their own children. So too did Kay:

> Now I feel myself that a good nanny – not a crèche – I don't believe in them – but somebody coming in can give them a pretty stable background. But, having said that, if I was having my own children . . .! I honestly feel that if you haven't enough money to stop work you should wait a few years before you have children. Definitely. I really would like to be at home with my children, because I know how impressionable they can be. I couldn't be like 'Diana'.

Clearly, then, and in spite of the counter models of femininity and mothering offered by their employers, this nanny labour force is characterised by an acceptance of the traditional social construct of motherhood in post-war Britain, and particularly by an acceptance of its emphasis on the importance of full-time childcare by the child's natural mother. It is the social practices of their own mothers, rather than those of their employers, which these women wish to reproduce when they, in turn, become mothers. Such ideas can only be explained in terms of emergent identities within which being a traditional wife and mother play the dominant part.

A further factor which undoubtedly proved instrumental in shaping the emergent identities of our respondent nannies was the presence within some of our nanny-producing households of much younger siblings. What we encountered here were situations in which the teenage daughter had been used as, and actively encouraged in, replacement childcare by her own mother. Here is Clare again:

> I used to watch Damon a lot for her. Just more or less taking him to A and B. And then bringing him back. Baby-sitting more or less I suppose! I used to enjoy it. I used to think, 'Oh this is great!'

Such activities were clearly seen at the time as responsible, as valuable in a supportive sense to their own mothers, as well as enjoyable. Once again, their potential influence in terms of identity formation is hard to assess, but it would seem that, in assisting their mothers in various ways with childcare as teenagers, the young women whom we interviewed began to identify in a positive way with at least some of the activities of mothering. For our respondents, the responsibility of temporary charge provided them with a definite sense of self-worth, and seems to have been associated closely with ideas of emerging adulthood.

As Clare suggested above, the temporary care of younger siblings is an activity similar to baby-sitting, although the social relations are very different. Located for the most part in another household within the neighbourhood, baby-sitting is carried out primarily for cash, and secondarily for love. It is, in short, one of the most obvious ways in which teenage girls can earn extra money, although most of our respondents claim to have done it because they liked children. The contradictions in such statements are fascinating, since for the most part baby-sitting as an activity involves sitting in a house whilst the children sleep, with very little in the way of interaction with children! Given this, we can only infer that what our nannies meant here was that it was the conceptual association between children and baby-sitting which they liked; an idea which resonates strongly with notions of responsibility and emergent femininity.

As might be anticipated, almost all of our respondent nannies had done baby-sitting frequently as teenagers. Here, for example, is Kay again:

> I'd always loved children. I mean I've baby-sat since I could pick a child up you know. And I was very close to a family nearby. Still am – still baby-sit for them in fact! And through my summer holidays when I was still at school, that was my full-time job. I went to help them out. And at weekends I baby-sat every Saturday night.

It is interesting to note in passing how baby-sitting, like cleaning, is yet another form of female 'off the cards' work located within the domestic sphere. Unlike other such activities, though, this is mainly a form of work constructed for teenage girls. In part the explanation for this gendering is one which has to recognise that baby-sitting is an activity which enables parents to be released for a short duration from childcare responsibilities. It therefore by definition is something which cannot, in general, be done by those with major parenting demands. But, in using the labour of 'responsible' teenage girls to baby-sit, parents are clearly not substituting any form of cheap replacement labour, but one which is considered suitable. Although unintentional, we would argue that the consequences here are similar to those which stem from the social practices discussed in relation to the parental home, namely an enabling of the development of traditional gender identities, in which mothering and childcare sit centrally.

The foregoing demonstrates that both the parental home and the neighbourhood are sites of specific social practices which have been at the very least facilitative of the development of traditional female gender identities amongst the young women whom we interviewed. These practices conveyed various deep-seated beliefs and associations about the relation between childcare and mothering (rather than, say, between either childcare and fathering or childcare and parenting) and actively enabled these young women to identify themselves with the care of young children. We are also in no doubt that all this was seen in highly positive terms by these young women, for – in undertaking temporary childcare responsibilities in both sites – they were able to construct themselves as valuable and valued participants in the social life of the parental home and the neighbourhood. This, as we now show, contrasts quite markedly with what they encountered at school.

School

Detailing the full educational experience of our respondent nannies is beyond our concern but three related issues are of importance. These are: the educational profiles of these girls; their attitudes to school; and the specific careers advice which they received. We will take each of these in turn.

The educational profiles of our two nanny labour forces are shown in Appendix 5.1 (5.1e and 5.1f). The two are identical and were corroborated by interviews with NNEB course tutors and surveys of students on NNEB courses. We have here a group of 16-plus school-leavers; girls who are either educationally average or below average. Such profiles were frequently, although not universally, accompanied by a range of broadly negative experiences and attitudes concerning school.

Thus, a sizeable minority, whilst not actively disliking school, essentially muddled through their school careers. Others encountered 'knocks' in terms of their supposed academic abilities, which clearly had affected their future aspirations. Both are illustrated below.

Clare and Louise are representative of the first group. Here Clare summarises her educational career:

INTERVIEWER: And did you leave when you were 16?
C: Yeah.
INTERVIEWER: So can you remember what you left with?
C: You mean me results?
INTERVIEWER: Yeah.
C: Just CSEs. It was a grade two English and a grade two childcare. Social studies I got a grade three. Just more or less roughly that. Maths I got a four. I was useless at that!! Really I wasn't very bright at school. It was all right you know. I just mingled in. Got on with it. But I was never quite bright enough I thought.

Louise (19), as the following demonstrates, had great difficulty in remembering her qualifications:

INTERVIEWER: And you left at 16?
L: Yeah.
INTERVIEWER: And did you enjoy it?
L: It was all right. Like I didn't mind it. I didn't hate it. But!! (laughs)
INTERVIEWER: So what qualifications did you leave with then?
L: I got three O levels and three CSEs. It was English, Art and uhmmm, you know I haven't got a clue! What was it? (thinks)
INTERVIEWER: Maths?
L: No!!
INTERVIEWER: Geography? Biology? Well it doesn't matter!
L: I think it might have been history. Something like that anyway!!

In Clare's case, we have an example of someone who, for whatever reasons, thought that she wasn't clever enough academically, and felt that this was demonstrated by her achievements – 'just CSEs'. For Louise, in contrast, school had evidently been something of a non-event! At the other end of the spectrum were those who had hoped to do well at 16-plus, but who didn't make the required grades. Stephanie (19) took nine O levels but passed just two and Kay found herself in much the same position.

In educational terms, girls such as those whom we interviewed – at least at this stage of their lives – are labelled as 'non-academic'. This evidently posed the teachers of our future nannies, as well as our future nannies themselves, some problems when it came to making choices at 16-plus. This was demonstrated starkly in the accounts which our respondents gave of careers interviews and counselling. For some, such as Rebecca, the question of 'what next?' was one great imponderable:

R: I didn't have a clue really when I was at school.
INTERVIEWER: Were you aware of what sort of options there were?
R: I knew that you could either stay on at school or you could go on to college or you could just go out and find a job. But I didn't really know what kind of job I wanted to do. I thought about being an infant teacher but I didn't have the qualifications when I left school to go on to college. So I thought of business studies. And then one of the teachers suggested this course (PCSC) and I thought, 'Well I'll do that and see how it turns out.'

For others, 'work experience' relating to childcare proved to be particularly formative. Thus for Kate (24), it was this which alerted her to the existence of childcare courses:

> The thing was I didn't know there was anything you could do to work with children. I only thought you could be a teacher. I didn't know there was a course for nursery nurses and that you could go and work in a nursery. I don't suppose I knew that I wanted to work with children. And at school we had two weeks when we had to do work experience. And I don't know why but I said that I wanted to do something with children. So they said, 'There's a nursery school at D.' So I said, 'Oh I'll go there.' And it was while I was there that I found out about the nursery course and it sounded quite good.

In other cases, the fact of saying in the context of a careers interview that they liked being with children was, it seems, sufficient to propel some of our respondents down the childcare course route. Karen (25) is illustrative of this situation:

> They didn't really suggest many things to you at school. They just said, 'What do you want to do?' And obviously, if it wasn't that outrageous – I mean if you said, 'I want to be a model' or something like that. But because it was down to earth – I said, 'Working with children' – they just accepted it. They didn't push you anywhere else. Maybe if I'd been given a few suggestions I might have done differently. I might have gone into catering. Or a hotel receptionist. Or, I don't know, I like fashions, I like working

behind the scenes. Those are things you know that I think I might like to have done.

The result in all three of these cases was that the girl in question ended up on a course which, in one form or another, involved childcare, and, at one level these cases illustrate the two principal routes by which teenage girls opt for childcare courses at 16-plus. For some of our respondents – the minority – this followed the explicit suggestion of professional advisers (specifically careers counsellors and teachers). However, for the vast majority the choice reflected either a general liking for children or a specific wish to work with them. In order to understand both routes it is essential to acknowledge the formative role played both by emergent adult gender identities and by traditional assumptions regarding 'a woman's place'. For most of the young women whom we interviewed, school had been a place of conspicuous under-achievement. By their own admission, these girls, for one reason or another, had not done well at school, and in some cases had failed to achieve what they had hoped for. In searching for their individual answers to the 'what next?' question at 16-plus, it is not surprising that those who made the positive decision of childcare for themselves, turned to an area which then occupied a prominent and positive position in their individual consciences, namely the responsibility of caring for young children. Here apparently was something which they liked doing and saw as valuable, and also as something which they felt competent to do, indeed able to do well – a strong comparison with the relative failures which they had experienced in the education system. For these girls, opting for a childcare (or related) course meant not just a chance to prove themselves a success, but the opportunity to succeed in an activity which could be seen as appropriate for a young woman. Indeed, we suggest 'failure' in one sphere to be the key which unlocks the conformist ideas which we saw being formulated in, and shaped by, some of the social practices of home and the neighbourhood. In contrast, for those young women whose route into a childcare (or related) course was more passive (illustrated above by Rebecca), it is the assumptions of professional advisers about supposedly 'non-academic' girls which prove to be critical. Here the question 'What can she do?' is answered by traditional expectations and assumptions concerning adult women's gender roles.

So far in our analysis of the production of our future labour force of nannies we have emphasised the importance of a particular emergent gender identity and suggested how some of the social practices associated with the parental home, the neighbourhood and school can be both productive and reproductive of such identities. We have also suggested that for 'non-academic' girls (and their careers advisers and teachers), a vocational training course involving childcare can be a relatively straight-forward, obvious, and even automatic, choice at 16-plus. However, such choices are frequently made in relation to, and in preference to, a range of other labour market possibilities. Indeed, we now show how traditional emergent identities enable the rejection of a range of other labour market and/or training options in favour of vocational training involving childcare.

Unpacking 'I'd always wanted to work with children'

By comparison with research examining the transition from school to the labour market and/or training for boys (Finn, 1987; Willis, 1979), the literature on girls is extremely limited. Indeed, Chris Griffin's (1985) study *Typical Girls?* remains the only work to focus explicitly on this issue (although see Cockburn, 1987). Whilst much of Griffin's research needs to be read, together with other work from the Centre for Contemporary Cultural Studies, as 'a female version of Willis' (see, for instance, McRobbie, 1978), her work suggests two points which have proved particularly useful to us. Firstly, Griffin points to the contraction in the youth labour market in the 1980s. Unlike in the previous decade, the transition from school to labour market in the 1980s was characterised by fewer 'real' jobs and by high levels of youth unemployment, masked by growing numbers on compulsory youth training schemes (Finn, 1987). For girls leaving school at 16-plus in this decade, the labour market options effectively boiled down to a choice between low-grade office/clerical work, retail sales, and a declining number of factory assembly jobs. Alternatively, they could opt for a vocational training course (for instance in catering or in hairdressing). Beyond this was compulsory Youth Training (YT). The second point which we pick up from Griffin is the way in which 'typical girls' think about future employment. Griffin's work demonstrates conclusively that 'non-academic' girls have definite perceptions of the sort of paid work available to them, and this was certainly something which we encountered in our research.

Our analysis of the transition from school to training for nannies had two components. First we surveyed two groups of students currently on NNEB courses in the north-east and south-east, asking them questions about the range of labour market and/or training options which they had considered at 16-plus, and to rank a series of employment and/or training options. Whilst this ranking exercise was not repeated with our interviewed labour force (because we felt it more appropriate only to discuss choices and decisions at 16-plus with them), it corroborates our interview findings. Tables 5.1 and 5.2 present the results of this survey. As is evident from Table 5.1, the majority of the sixty-eight students responding to our questionnaire *only* considered childcare options at

155

Table 5.1 Labour market and/or training options considered at 16+ by students on NNEB courses in case study areas

Options considered	North-east No. (n = 34)	Percentage	South-east No. (n = 34)	Percentage
Childcare/childcare related training course (e.g., NNEB, PCSC) *only*	16	50	21	62
Any other form of 'care' training (e.g., nursing, physiotherapy)	7	22	6	19
Staying on at school/ college	4	13	1	3
Hairdressing	2	6	–	–
Clerical/secretarial work	–	–	2	6
Other (catering; fashions; dance; air hostess; teaching; YT; design)	6	19	7	22

Source: Fieldwork data
Note: Since some respondents considered >1 option, numbers do not total 100 per cent.

Table 5.2 Employment and/or training rankings by students on NNEB courses in case study areas*

Employment/training options	North-east Percentage maximum score	South-east Percentage maximum score
NNEB	93	81
Other 'care' course (e.g. PCSC)	65	52
Nursing	77	59
Secretarial/clerical	68	44
Factory assembly	35	24
Retail sales assistant	48	41
Youth Training	32	32

Source: Fieldwork data
Note: * Rankings ranged from 5 (highly desirable) through to 1 (highly undesirable).

16-plus.[6] Those remaining actively considered a small range of other options (for example, hairdressing, catering and secretarial work). A second question asked students to score a range of employment and/or training options on a scale of five (highly desirable) to one (highly undesirable). The choice of options reflected the labour market and/or training options generally open to 'non-academic' girls at 16-plus. Obviously an inbuilt bias existed here, in the sense that the NNEB option was highly rated by NNEB students! But what is more interesting is the marked distinction drawn by these students between the broadly care-related options (NNEB, a care course and nursing) and the remainder (particularly factory assembly and YT schemes). This suggests that care-related occupations and/or forms of vocational training are seen most positively by the types of girls who enrol on NNEB courses.

Such views accord strongly with the types of emergent gender identities which we have been emphasising throughout this section. Indeed, we maintain that they have to be understood, at least partially, in terms of such identities. Seeing oneself predominantly as a future full-time wife and mother means that it is the activities associated with caring for others which will be seen to be of greater value than those occupations without such associations and/or characteristics (for example, clerical work, factory assembly and retail sales), some of which appeal to very different constructions of femininity. Moreover, we suggest that identities such as these are critical to understanding why the majority of those girls surveyed *only* considered a childcare course at 16-plus. Both in wanting to be and in constructing themselves as future full-time wives and mothers, girls such as these find themselves in very difficult circumstances at 16-plus. However much we wanted to resist such notions, we have no doubt that these girls' emergent identities allowed them to see their own labour force participation as either temporary or secondary, and as filling in time before the real work of marriage and motherhood began. Undoubtedly, for such girls, finding vocational training courses which focused on certain of the activities which they could envisage themselves doing in the immediate future solved the problem of 'what next?' at 16-plus. In a sense they didn't have to consider anything else because this was something which 'seemed OK'; it coincided with their own images of their future lives and it was something which others (mothers, fathers, friends and teachers) seemed to see as appropriate for them. Moreover, in that training with a qualification was being offered, we suggest that this route would have appealed more strongly to girls from predominantly white collar backgrounds than a 'dead-end' job in either a shop or an office.

Such findings, and their suggestions as to the proactive role played by emergent gender identities, were supported by our interviews with young women now nannying. They appeared to have evaluated, in an identical way to those surveyed, the labour market and/or training options open to them. Moreover, the majority had not considered anything other than a vocational training course involving childcare. Anna (20) provides a good illustration of the way in which a traditional emergent gender identity proved critical in enabling someone who could have been a secretary to opt for an NNEB course. Anna left school with five O level passes and distinctions in stage one typing and secretarial examinations. Here she tries to grapple with the issue of why she chose to work with children, rather than become a secretary:

157

A: I put down that I wanted to work with children on my careers form. I could have put secretary down but I didn't, though I did typing.

INTERVIEWER: Why didn't you want to do typing?

A: Because I wasn't interested. Not like looking after children.

INTERVIEWER: So you didn't fancy working in an office?

A: I wouldn't have minded it, but to me it wouldn't have been as satisfactory as working with children, with job satisfaction.

INTERVIEWER: What do you think that satisfaction is?

A: I don't know but when you're with children and they do something, you feel proud that you've helped them. And then there's the friendship. You develop a friendship.

For Anna (although she – like all of our respondents – found it difficult to get beyond the standard 'I'd always wanted to work with children' response to our questioning) working with children offered the chance to develop, to help and to care for and about a specific someone else. It was, in other words, the emotional associations of the work and its social relations which proved attractive to her – far more attractive than anything which a word processor could offer! So, it is caring (in the sense of both the physical activity and the emotional bonds which the activity of caring both produces and manifests) which proved critical in shaping her individual decisions at 16-plus. We consider these ideas and associations to have been of major importance in structuring not just Anna's decisions but the decisions of many of the young women whom we interviewed. For them, the opportunity to care for and to care about young children was not just attractive, but central to their entire emergent view of their adult selves.

FROM CHILDCARE COURSES TO NANNYING: FROM TRAINING TO THE LABOUR MARKET IN THE CHILDCARE SECTOR IN THE 1980s

Having established how and why the young women whom we interviewed came to opt for vocational training courses concerned with childcare, we now consider the processes through which they came to take jobs as nannies.

There are five main forms of work in the childcare sector: within day care for pre-school-age children (i.e., in local authority, private or workplace nurseries); in primary education (in relation to rising 5s); in hospitals (in specialist children's wards); as registered (and unregistered) childminders; and as either a day or live-in nanny within a private household. These forms of work are mediated by qualifications and the female life-cycle. Thus, whereas holders of the NNEB certificate can, in theory, work in all five areas, those without this qualification are restricted to childminding and nannying. The NNEB certificate therefore acts as a qualifications barrier; it restricts access to the career structure within nursery work. By contrast, it is the specific stages in the female life-cycle which both facilitate childminding for certain women and prohibit other women from working as registered childminders. Women with young children constitute the majority of registered childminders in contemporary Britain.[7] Such characteristics reflect the compatibility between this occupation and the demands of a family. Childminding, however, by definition requires one to have a home from which to work, effectively restricting this form of employment to women who are either owner-occupiers or council tenants. In other words, it is highly unlikely that young women wishing to work in childcare (but who live in the parental home) could work in such a capacity.

Nationally some 42 per cent of all those qualifying from NNEB courses in any one year (approximately 5,000–6,000) take up employment as nannies.[8] Nannying therefore constitutes the largest single occupational destination for NNEB holders. By comparison, approximately 21 per cent go on to work for local education authorities; 6 per cent for local authorities (in public sector day care); 6 per cent for health authorities; 10 per cent for private and/or voluntary organisations; 5 per cent take further training and around 3 per cent move out of childcare altogether.[9] Moreover, our survey of young women currently on NNEB courses confirmed the popularity of nannying.[10] Although statistics are unavailable on labour force participation for those holding the Preliminary Certificate in Social Care, it would seem fair to suggest that, for those opting to work with children, nannying constitutes one of the few employment openings.

The above shows clearly that, once having taken a vocational training course concerned with childcare, the majority of young women enter the labour force as nannies. For some, the NNEB qualifications barrier means that this has to be so, but for NNEB holders this is not the case. So why is it that the NNEB holders in particular do so? We can answer this question in two ways – in relation to the form of childcare provision in contemporary Britain and in terms of the potential attractions which young women see in nannying. As will become clear, we consider the first of these to be important.

The structure of employment in childcare is such that for much of the post-war period, and increasingly through the 1980s, jobs as nursery workers in local authorities (LAs), in local education authorities (LEAs) and in health authorities (HAs) were increasingly hard to come by (Cohen, 1988). The expansion in private sector day care in the late 1980s, as well as in workplace nurseries, is argued to have changed this position somewhat.[11] However, in the recessionary context of the early 1990s, such observations remain speculative rather than demonstrated. A number of the nannies with the NNEB qualification whom we interviewed made reference to both the perceived and real difficulties in finding work in local authorities, local education authorities and health authorities during the 1980s. Here, for example, are Sophie (23) and Liz (24):

INTERVIEWER: So why did you choose nannying?
S: Well basically because it's so hard to get a job in a school. You need the experience before you can get into a school. So the only thing to do was nannying.
L: If a school vacancy had been there I'd have taken that. But I needed a job straight away when I left the tech.
INTERVIEWER: So the pressure was on to take a nanny job?
S: That's right. There were just no vacancies at all.

Others who made similar points were Kate (24) and Karen (25):

Kate: What I'd always wanted to do was to work in a special care unit with premature babies. We did that at college and I really, really enjoyed it. So I wrote away to all the hospitals but there were no jobs at all. I'm on the waiting list, so that's as much as I can do really. That's my main goal but I know that realistically there's 'no way'.

Karen: I liked the hospital but I found that it was very difficult to get in. There's only two nursery nurses or something like that to a ward and they – once they get in a job, they don't leave. So I sort of put that to the back of me mind.

By comparison, jobs as nannies – especially for girls with the NNEB certificate – are relatively easy to come by. Not only are such vacancies placed by parents in local newspapers, in *The Lady*, with Job Centres and with employment agencies, but many parents go direct to NNEB course tutors with their vacancies. We analyse this situation in more depth in Part Two, but for our purposes here it is important to appreciate that such actions mean that the colleges themselves act as informal job centres. In this respect, the comments of NNEB tutors are enlightening:

INTERVIEWER: And do you get people ringing in?
TUTOR: Oh yes. I've got one now. She's a lecturer. Her nanny is leaving and she wanted somebody now, but she's prepared to wait to get the best of the bunch. People are always ringing in. I then read it out to the year group and put it on the noticeboard. Those that are interested apply and then they're interviewed and so on. Most are fixed up by the time they leave.

TUTOR: We don't actually run an agency as such but if it were a school looking to replace a nursery assistant, usually they would phone us and say, 'Have you got anybody suitable?' And that tends to happen with the day nurseries as well. Just a phone call. . . . For private nannies, people phone up . . . and we have a system whereby we send out a form to prospective employers which we then put on the noticeboard and students can take it from there. I try to weed out people who sound as if they're looking for mothers' helps or are thinking of £40 a week or whatever.

One of the consequences of such actions is that young women on NNEB courses often possess privileged local knowledge of nannying vacancies within the local labour market. Moreover, given tutor screening, these vacancies are quite likely to be amongst the better of such jobs on offer. What we have here then is a closed and informal labour market for NNEB students. Here is a form of training which delivers – albeit unintentionally and informally – a job. In such circumstances, and given the pressure on young people to get a 'real job', it is not difficult to see how the majority of NNEB students end up nannying, and why the nannies whom we interviewed with the NNEB certificate found it relatively easy to enter the labour market in this way.

Both the ease of the transition from training to working as a nanny for NNEB students and the workings of the informal labour market for NNEB students are illustrated in the following two extracts:

Louise: And I just got a job nannying. I don't know how. They kept on coming into the college – you know how they write in. And I just took all the names and addresses down. And this one came in. And I thought, 'Well it's in X, it's dead close. That'll be good. I'll go for that'.

Kate: I don't think I actually decided I wanted to become a nanny at all. But towards the end of the course I was really lucky in that somebody rang in with a nanny job with a six-week-old baby and I thought, 'Well I'll go for that.' That more or less settled it.

The reality of employment for childcare workers in Britain in the 1980s, and the resurgence in demand for the nannying form of waged domestic labour during the same period, together shaped the transition from training to the labour force for many NNEB students in the 1980s. However, at the same time, there is little doubt that this path was legitimated ideologically by the particular characteristics which working as a nanny appear to offer. Thus, for a number of the NNEB certificate holders whom we interviewed, nannying had seemed attractive, precisely because it presented the opportunity to care for a young baby, the

161

chance to develop one-to-one relationships with children and because it seemed to offer a form of freedom – in short the opportunity to 'be your own boss'. Anna encapsulates this situation:

INTERVIEWER: So why was it that you wanted to do nannying?
A: I wanted to be my own boss, I wanted to do things with the children. I wanted to work out my own ideas; see how I got on with them. And then you only have the one person to look after. It's yours to look after.

Similar ideas were presented by the nannies whom we interviewed who did not have the NNEB certificate, but in these cases such ideas undoubtedly worked to obscure the restricted form of their labour force participation, rather than to mask the dearth of career-structured openings available to young women with a professional childcare qualification. Furthermore, and as might be anticipated given both the preponderance of NNEBs working as nannies and the existence of informal job centres for NNEB students, the young women without the NNEB qualification whom we interviewed had found it initially more difficult to find employment. Moreover, some of these had also encountered particularly poor and/or difficult employment conditions (i.e., they found themselves in the sorts of jobs which course tutors screened out as 'unsuitable' for their students). Janice, for example, experienced both these problems:

J: When I left college (PCSC) I worked in a restaurant full-time and that's how I got my first job. I mean I went for hundreds of interviews.
INTERVIEWER: Nanny interviews?
J: Yes. One was in X. The lady asked me to go through, and I would have had to have started at 7.30 a.m. She wanted me to go on holiday with them and keep the house and so on. . . . Then, while I was working in the restaurant, Rick came in and said that 'Anna' had been ill and would I like to go to work for them. So I said, 'Yes'. And that's how I got in.
INTERVIEWER: Can you estimate how many interviews you went to?
J: God! Maybe twenty.
INTERVIEWER: Did you find not having the NNEB a handicap?
J: Sometimes, yes. As long as you went in with the attitude that you could work with them and that you were no less a person than an NNEB, then you had a chance. But, I mean it's like all jobs. Unless you've got experience you know, they wouldn't entertain you. But if you had the NNEB they'd say 'Oh well'.

The foregoing demonstrates that, once having been through vocational training involving childcare, young women find themselves confronted by the realities of childcare provision in contemporary Britain. The paucity of day-care facilities, together with central government imposed cuts in LA, LEA and HA expenditure, have severely restricted employment within specific sectors of childcare and made access to these forms of employment particularly difficult for newly qualified women. Moreover, the current recession means that this situation is hardly offset by a surfeit of jobs in private sector day care. Effectively, then, qualified young women wishing to work with children in contemporary Britain have had little choice but to be a nanny. Much as the form of childcare provision in contemporary Britain, together with the ideology which underpins this, have led to a resurgence in demand for nannies by households with both partners in service class occupations, the same context can be seen to have produced a labour force of nannies in the 1980s, rather than a labour force of nannies, nursery workers and children's nurses. Not only do the young women emerging each year with either NNEB or PCSC qualifications constitute a ready supply of childcare workers, but – if they are to practise the childcare training which they have received – a growing percentage of them in the 1980s would appear to have had no choice but to do so as the nannying form of waged domestic labour. In short, the nannying labour force in contemporary Britain is one of displaced professionals. It is this pool of young women which the new middle class has been able to draw on to satisfy its increasing levels of demand for childcare-related waged domestic labour in Britain through the 1980s.

The situation outlined above is also significant as regards labour supply. With over 2,000 NNEB students per annum known to be entering the labour force as nannies, and an unknown number of PCSC holders (as well as unqualified women) working in this capacity, there is a steady (and possibly slowly expanding) pool of young women willing to work as this particular form of waged domestic labour. Moreover, as is suggested by Appendix 5.1g, the majority of these women work in this capacity for a period of years. Given this, it is clear that there is currently no shortage of women available to satisfy expanding levels of demand for waged domestic labour for childcare (although, given the geographical distribution of the middle classes, parents may encounter difficulties in finding 'qualified' labour in sufficient quantity in particular areas of the country, particularly London and the south-east). However, at the same time, it needs to be recognised that the availability of this pool of qualified childcare labour is itself, at least in part, a reflection of the paucity of collective childcare provision in contemporary Britain. Should there be a major expansion in either private or public sector daycare provision in Britain in the future, we anticipate that a sizeable proportion of those NNEB holders currently working as nannies, or women undergoing training, would opt for this form of employment. Undoubtedly, such circumstances would spell major problems in labour supply for those middle-class households who continued to wish to employ a nanny in relation to their childcare needs.

CONCLUSIONS

In this chapter we have examined the conditions and circumstances which have been critical in the production and reproduction of a ready supply of cleaners and nannies in Britain throughout the 1980s. We have argued that, whilst the supply of private domestic cleaning labour still contains older working-class women supplementing the state pension, and working-class mothers fitting in the odd bit of informal work around the demands of children, it also contains growing numbers of benefit-dependent women. Given the structure of the benefit system, such women have few alternatives but to work as private domestic cleaners in contemporary Britain. We consider these women to have played a major part in enabling middle-class households to satisfy their expanding levels of demand for cleaners through the 1980s. As regards the nanny, we have shown how both the production of a steady annual supply of qualified childcare workers and the context of day-care provision in contemporary Britain combine to make available to middle-class households a ready supply of qualified professional workers, many of whom have little choice but to work as nannies. Whilst there is no doubt that the processes and trends which have generated these particular forms of waged domestic labour in contemporary Britain have been critical to their aggregate level reproduction, the social relations of both occupations play a major part in their reproduction within individual middle-class households. We turn now to an examination of these particular social relations.

Part II

NANNY AND CLEANER EMPLOYMENT WITHIN CONTEMPORARY BRITAIN

INTRODUCTION

In the second part of this volume we examine the two most important forms of waged domestic labour in contemporary Britain, the nanny and the cleaner, as they occur within dual-career middle-class households, looking specifically at their organisation and social relations. Chapter Six is concerned with the nanny; Chapter Seven with the cleaner. Whilst there are similarities between the two forms of waged domestic labour, some critical differences exist, both in terms of the organisation of the work and their social relations. We consider these differences to reflect the form of labour substitution involved in both cases, as well as the varying ideological constructs which suffuse and shape both forms of waged domestic labour. In more general terms, our work can be read as extending Graham's critique of the feminist concept of caring (Graham, 1991) By definition, both the forms of waged domestic labour with which we are concerned are home-based, as well as care-related. Both involve wage–labour relations. But both are also characterised by social relations grounded in false kinship and/or close friendship. When looking at forms of care-based work (such as waged domestic labour), we need to avoid conceptualisations based on assumptions which equate care with home-based, unwaged labour shaped by the social relations of kin. Instead, care-based work (and caring) needs to be recognised as contingently related to location, and as a form of work which can exhibit the social relations of both wage labour and kinship (false kinship) – in some cases (as in certain forms of waged domestic labour) in conjunction with one another.

6

NANNY EMPLOYMENT IN CONTEMPORARY BRITAIN

As we showed in Chapter Two, the nanny constitutes one of the two most important categories of waged domestic labour in contemporary Britain, and the category of waged domestic labour for which demand increased most significantly through the 1980s. In this chapter, and in the spirit of the second part of this volume, we switch from accounting for this to examining nanny employment as it occurs within individual dual-career middle-class households. The chapter is divided into three main sections. In the first we outline and illustrate the basic character-istics and forms of nanny employment, looking specifically at the tasks, activities and typical days of nannies at work in Britain in the late 1980s and early 1990s. Moving on from this, we focus on the two major contradictions of contemporary nanny employment; namely, the con-struction of the nanny as, at one and the same time, a childcare professional and a mother substitute, and the construction of the social relations of nanny employment in terms of both waged labour and false kinship. Finally, we concentrate on accounting for the form taken by these social relations.

At this juncture we should add that we do not wish to extend this representation to encompass the entire UK. Indeed, there are two lines of differentiation which we would expect to emerge from further analysis. The first of these is the form of nanny employment within London. As we have already seen in Chapter Two, London constitutes the most important concentration of nanny employment in Britain. Moreover, and unlike our study areas, it is dominated by live-in forms of employment, as well as by frequent job turnover. We anticipate that this situation will have a profound and limiting effect on the develop-ment of social relations structured on predominantly false kinship lines. A second qualification pertains to class. Nannies – as we argued in Part One – have long been part of the lifestyle of the British upper classes, aristocracy and royalty. We do not anticipate that the characteristics and relations which we consider here would extend to this different class context.

INTRODUCING NANNIES AND THEIR EMPLOYERS

We encountered five forms of nanny employment in the course of our fieldwork. The two simplest were those in which one nanny worked for one household, in either a live-in or live-out capacity. Although the second of these arrangements proved to be the most common practice (Appendix 5.1), this is not the case within London (where live-in forms of employment dominate).[1] Moreover, even within our fieldwork areas, more complex forms of nanny employment existed. The first of these is the nanny-share – a set-up in which one nanny is shared (pretty much equally) among households (usually two or three). In these circumstances the households are usually located near to one another, and are frequently close friends and/or neighbours. Nanny-shares can involve either live-in or live-out arrangements and involve the use of all houses within the employer network. The final, and most complex, form of nanny employment encountered was that where a nanny is employed in the main by one household but has a number of secondary employers, all of whom use the nanny for only a few hours a week. In such circumstances, although the bulk of a nanny's weekly earnings come from the primary employers, this is 'topped-up' by casual payments (usually on an hourly or session basis) from the others. Given that this pattern of working is based exclusively in the home of the primary employing household, such arrangements involve both childcare in the parental home and care located outside it. This form of employment is more correctly labelled as involving a nanny (childminder). The following individual cases provide examples of all five forms of nanny employment.

The Lyles and Diana (live-out, one employing household)

Diana is 21 and has been the Lyles's day nanny for twelve months. She works Monday to Friday from 8.15a.m. to 6p.m., looking after the two children, Natalie (6) and Adam (4). Diana is single and lives with her parents a few miles from the Lyles. Diana's father is a sales representative, and her mother is a lunch-time controller and part-time cleaner at a local school. Diana took the NNEB qualification at the local technical college and this job is her third nanny job since leaving college. She is paid £150 per week gross. Both Diana's employers work full-time. Stephanie Lyle is 37, a biochemist and a director of a multinational chemical company. James Lyle is 41 and a self-employed architect. Both partners were born and brought up outside the Reading area. However, they have lived in this area now for nearly fifteen years.

The Maxwells and Janet (live-in, one employing household)

Janet is 22 and single. Her father is a television engineer and her mother a housewife. She has been the Maxwells' live-in nanny for three years (i.e., since Rebecca, their eldest child, was a baby). The Maxwells also have a young son (Thomas, aged six months). They pay Janet £95 per week gross. Working for them is Janet's first job since leaving college in South Wales with the NNEB qualification (she was advised of the vacancy by her course tutor, whom Corinne Maxwell had approached directly). Corinne Maxwell is herself from South Wales and is a lawyer. Her husband is a scientist. Both partners are in their late thirties and moved into Berkshire over five years ago.

The Newells and the Orwins and Tracey (live out, nanny-share)

Tracey is 22 and has been married to Darren (a carpenter) for five months. Tracey took her NNEB qualification at the local technical college. She works five days a week from 8a.m. till 6p.m. for the Newells and the Orwins, and is paid £125 per week gross. She looks after her 'charges' – 4-year-old Daniel (Newell), and 3-year-old Alexander (Orwin) in alternate homes each week, and has been working in this job for a year. Peter and Josie Newell are both in their mid thirties and have lived in the Newcastle area for nearly ten years. He is a police constable; she works in insurance and her job means that she regularly has to work outside the local area. The Orwins are also in their mid thirties. Keith Orwin is a university research worker and Hazel Orwin a computer programmer. Both the Newells and the Orwins moved to the Newcastle area from the south-east region.

The Peters and Kay (live-out, one primary employing household plus secondary 'top-ups')

Kay is 23. She lives with her parents in a village outside Durham City. Her father is a sales representative, and her mother a secretary. Kay took her NNEB qualification at the local technical college and, since leaving college, has had three nanny jobs. The first two were in London and lasted for three and three and a half months respectively. She then returned to the north-east to a job which was initially 'sole charge' for one employing household – the Peters. She has worked for the Peters for over three years. However, recently her job has changed to involve additional working for a secondary employer. In addition to her two main charges – Harry (3) and Laura (18 months), she now cares for Justin (6 months) in the Peters' home on two afternoons a week

Currently Kay earns £90 per week cash; a wage which is made up of

£80 from the Peters and £10 from Justin's parents (she is paid £5 per session for his care)

Both Mr and Mrs Peters are in their late thirties. Diana Peters is a university administrator and Robin Peters a head chef. Justin's parents are a university lecturer and a part-time university research worker, both of whom are in their mid thirties. All four parents are from outside the north-east region.

The Quigleys and Jennifer (live-in, one primary employing household plus secondary 'top-ups')

Jennifer is 24 and single. Both Jennifer's parents are public school teachers. She has no formal childcare qualification but worked previously as a care assistant in the boarding school where her father is a house master. Jennifer currently lives in with her primary employers – the Quigleys – and their two children, Stephen (6) and Jonathan (4). The children are now both at school but, since the family still require after-school and holiday care for them, they have come to a complex arrangement with some close friends and neighbours who have a 2-year-old daughter ('Keia'). Jennifer continues to live with the Quigleys but, during school hours, she cares for Keia on her own. Jennifer's weekly wage is £146 gross per week, and is made up of £96 from the Quigleys and £50 from their friends. Sheila Quigley is a teacher and Andrew Quigley a research scientist. Both are in their late thirties and moved into Berkshire following higher education.

Such constitute the bare bones of nanny employment. What of the day-to-day reality? Here we detail the day-to-day activities of the five nannies introduced above.

Diana

D: I get here; they are both up, not dressed. Give Adam his breakfast, because he doesn't usually have it till I get here. Natalie goes and gets dressed. Dress Adam, clean his teeth. Take Natalie to school, go on and take Adam to nursery, then do something or other. Sometimes a walk, sometimes the shopping, depending on the day. Pick Adam up at twelve. He watches his programme on the television from twelve to twelve-thirty. Cook his lunch. We usually listen to the news while we eat lunch at one. Stack the dishwasher. We both watch 'Neighbours' everyday, then we usually do jigsaws or something – an educational something – or Adam has people to play. So it all depends. Thursdays and Fridays we usually have

171

<table>
<tr><td></td><td>people to play all afternoon and then he goes off and plays with them, either upstairs or in the dining room, and I talk to the nanny. So that's a typical day . . .</td></tr>
</table>

people to play all afternoon and then he goes off and plays with them, either upstairs or in the dining room, and I talk to the nanny. So that's a typical day . . .

INTERVIEWER: So do you pick Natalie up from school?

D: Natalie I pick up at three and we get back here at half-past. They have juice and a packet of crisps and they watch the children's television as long as it's been okayed by me. Then mummy comes in and I go home after saying, you know, 'He's been a brat today' or 'He or she's been a so and so today' – or whatever.

Janet

I get up at half past seven and now Rebecca and I go to play group in the morning. So we get up, get washed and dressed. Breakfast is over with and she has to be in nursery for half-nine. I drop her off at the nursery and come back and do whatever I've got to do – ironing, washing. And then I go and pick her up and then usually we see somebody. We maybe go to somebody's house for lunch, or they come back here for lunch and she plays with her friends. We get back here for half-four or five. When she's had her tea she has a shower or a bath, and then 'mum' takes over.

Tracey

Well I'll do it from Alex's house, because that's where I've been this week. I usually get there for between quarter-to-eight and eight and Daniel is dropped off at five-past. When I arrive Alex is usually still in bed. I think it takes 'mum' all her time to get herself sorted out in the morning! Anyway, I get there, have a cup of tea and get Alex up. I wash and dress him straight away because I don't want to be doing that when Daniel arrives. Then I brush his teeth and Daniel plays a game or reads a book or something. I have to take Daniel to school in the mornings and that starts at ten-to-nine, so we are always out of the house by twenty-to. I have to pick him up at half-twelve. What Alex and I usually do then is to go round and see another nanny and he plays with the children while we have a natter. Sometimes we go home and I do something with him on his own like teaching him to write his name or we watch telly or something like that. Sometimes we go for a walk. At quarter-past-twelve we leave wherever we are and go and pick up Daniel from school. Then we go home for lunch. If the weather's fine we might go out for the afternoon. More often than not we go to see another nanny. We spend a lot of time out of the house.

That's more interesting for me and the children. If the weather's too bad we stay in and do something crafty or something like that. I let them watch television sometimes as well. I have to do the washing and ironing and keep the place tidy where we've been and, really, plonking them in front of the television is the only way to keep them from under my feet when I want to get on with something. Then about four forty-five I start to get the evening meal ready. Then I bath both children, even if they're not in the right house. And then whichever mum or dad is doing the picking up – it's usually the mum – they come at quarter-to-six and I leave at six.

Kay

Get here for half-eight. They're usually still in the middle of breakfast! And then Diana usually rushes off for a shower and I take over. Robin's usually gone by then. Then – when Diana's gone – I'll bath them both, get them dressed – whatever. Then we go out. Sometimes it's to another nanny's. Tuesdays Harry goes to play group and Laura and I'll go to another nanny's. She'll play and we'll chat. Usually we'll have lunch at the other nanny's – or if they come to us, lunch is with us. Then we come home and Laura has a sleep and I'll do something with Harry – drawing, painting or something like that. And then, when Laura wakes up I'll start doing the tea. Robin's in for four but I finish at five-thirty when Diana gets back. She asks me what they've done, if they've eaten and what have you and then I go.

Jennifer

It's sole charge. Get them up in the morning, take them to school and in the evening collecting them. Washing – basically the boys' clothing and bedding. I do a bit of ironing – just the general things. I don't do the cleaning or anything like that. I start at seven-forty-five and finish when Sheila gets home. Officially she gets home at five-thirty but there are quite a lot of times when she gets home before that. When she comes in I can go if I want to. Today they were already up. I got the breakfast, washed up and got them organised for the day. Took them to school. Came back and did a bit of clearing up. Keia had her lunch and I tidied up and took her up to nursery and then I had the afternoon off. At three-ten I collected the boys from school and they went to a friend's to play. Then we came back and I cooked dinner. It's very difficult to say exactly what you do. It doesn't sound a lot but with small children involved it takes you most of the day.

173

The 'typical day' for nannies is a long one (often between nine and ten hours) involving sole charge of at least one child (and often as many as four). The day usually begins with getting the children washed, dressed and fed, and (in the case of older children) taking them to school and/or nursery. Beyond this, day-to-day activities generally include performing at least some, and in a number of cases (usually where the nanny lives in) all of those reproductive tasks necessary to the maintenance of the children (for example, children's washing and ironing; changing their beds and washing their sheets; tidying children's bedrooms and toy areas; cooking lunches and teas; and getting children ready for bed). In between times the nanny carries out a range of other home-centred activities with the children – for instance, encouraging play and instructing in arts and/or crafts, painting, writing and cooking. In almost all the circumstances we encountered, this daily routine was punctuated by regular visits to child-centred activities such as 'Bounce' and 'Tumble Tots', as well by visits to nanny/mother and toddler groups, playgroups and other nannies.[2] Both toddlers' groups and visits to other nannies were as important facets of day-to-day routines as the basic reproductive tasks and other activities. Indeed, these provided key social contact points – for nannies as much as for children.

Although we have emphasised routine and generality here, it is important to stress that these daily practices are mediated by the age of charges, by employer expectations and by the particular form of nanny employment. For example, the basic reproductive tasks occupied most of the time of those nannies caring for young babies. In contrast, for those with older pre-school-age charges other activities took priority, although fitting these in around the structure of employers' days and basic reproductive tasks could prove problematical. Regardless of their nature, employers' expectations also proved extremely important influences on 'typical days'. Indeed, whether making either minimal or considerable demands regarding reproductive tasks, employers were shaping the possibilities for nannies in the sphere of other activities. Finally, the form of nanny employment was highly influential. Thus, whilst those nannies working in a live-out capacity for one employing household had considerable potential control over day-to-day activities, those working either in nanny-share arrangements or for primary and secondary employers were far more restricted – whether by absolute numbers of children in their care, or by the differential time–space structures of employers. Nevertheless, variation apart, it is basic reproductive tasks and other activities which constitute the key day-to-day activities performed by the nanny. Given this, there is considerable similarity between the 'typical day' of the nanny and that of mothers at home with young children. And yet, in spite of the similarities, there are important differences between the work of the 'housewife' and that of the nanny. Punctuating accounts such as those of Oakley's 'Patricia Andrews' of a 'typical day' are references to specific domestic tasks associated with house-care, i.e., cleaning floors, 'doing' the rooms and washing the 'nets' (Oakley, 1974). She also spends a considerable portion of each week washing and ironing. That this is so is unexceptional: Patricia Andrews' work is that of a 'housewife'. It therefore combines the activities of daily household social reproduction and childcare. The nanny – at least in theory – is *not* a housewife but the reality is somewhat different. As we now show, it is precisely the existence of the full-time mother/housewife – albeit in reduced numbers – which generates one of the major contradictions to be found within nanny employment in contemporary Britain.

MOTHER SUBSTITUTES – CHILDCARE PROFESSIONALS

Throughout our fieldwork one of the contradictions which we returned to repeatedly was the elision between the nanny as a childcare professional and the nanny as a mother substitute. For both nannies and employers, the reality of nanny employment was that it involved, at one and the same time, the employment of someone who was – for the most part – a qualified and/or experienced childcare worker, but whose work bore strong similarities to the unwaged work of the housewife and mother. This tension was revealed both by parents and by nannies. We begin our exploration of this issue with the comments of the Maxwells.

In explaining how and why they came to be employing a nanny, Paul Maxwell remarked:

> There were various options. Au pair, mother's help and nanny. Of the three the au pair is cheap but she is never going to stay for more than a few months at a time, is not going to be trained and is probably not even capable of looking after children. The mother's help is an in-between. And the nanny is the best of the three. *A trained professional* [our emphasis]. The most expensive. In it for a career.

This was also the view of Corinne Maxwell:

> They have always been very young children on both occasions. Six months and ten weeks and we didn't mind leaving them with a strange person, as long as we thought they could cope. *As far as I was concerned we wanted somebody trained* [our emphasis].

The Maxwells' wish to employ a trained professional to care for their children (something which was echoed in a considerable number of those households whom we interviewed) transfers through to their expectations of Janet (their nanny). As Paul and Corinne make clear,

Janet is expected to take full responsibility for the basic reproductive tasks associated with childcare, as well as the educational and social development of Rebecca and Thomas.

PM: We just said, 'You are responsible for the children'. Full stop. So the children's play area, bedrooms, the bit of carpet underneath where they eat, all are kept clean and tidy by the nanny. She is not responsible for hoovering and dusting the non-child areas. Washing children's clothes, preparing children's food, washing children's cutlery and crockery. They are all her responsibilities. Making sure the pushchair's in working order. Not necessarily getting down there with a spanner, but telling me if it wasn't working. Everything to do with the children is her responsibility. Everything to do with adults is ours.

CM: She has to make sure they are clean, fed and looked after if they are not well. She has to teach them discipline and also educational things – sitting down with a book or sitting down with them and playing games. . . . I came home one day and the whole house was labelled, the chairs, the fridge, everything.

However, and in almost the same breath, the Maxwells describe Janet's work thus:

INTERVIEWER: What do you ask your nanny to do? On a day-to-day basis, what are her responsibilities?

PM: *To look after our children as we would during the day* [our emphasis].

CM: To do things that are fun and busy. Plan things. Take them swimming. Wednesday is nanny and toddler day and then they all go to lunch somewhere. Thursday is ballet in the afternoon for Rebecca. Friday they usually have friends to play or visit someone. . . . *We have always established ground rules. Janet works within our guidelines and does not do anything of which we disapprove. So if we say, 'You should never smack them', she will never smack them. She will never do anything of which we disapprove. . . . I am confident enough that Janet is professional enough to say, 'You tell me how you want it done and I'll do it your way'* [our emphasis].

For the Maxwells then, Janet's trained professional status is something which, in practice, is subordinate to their ideas on childcare and development. She is, in effect, a substitute for parental care, and specifically for the type of care which Corinne Maxwell would have provided.

The Maxwells are an interesting case, not least because, whilst Janet is clearly perceived as a subordinate party within this arrangement,

Corinne Maxwell's comments allude to the centrality of trust, consensus and negotiating frameworks to her employment. As can be seen from the following two instances, the same three structuring principles are the key to the daily reproduction of the nanny form of waged domestic labour within individual households. The two cases are ones which we have met before, in Chapter Four.

The Joneses

Janet and Peter Jones are the media couple whom we introduced in Chapter Four. They have a daughter aged $2\frac{1}{2}$. Since she was born they have employed two nannies, the first of whom Janet Jones labelled as 'a disaster'. The second has been with the household for two years and is a clear success. It is instructive to record Janet Jones's comments on the first experience and counterpose these with her representation of the development of the current situation.

INTERVIEWER: You don't have to go into great details about it, but what was the problem?

JJ: She was very young. She appeared to be confident and capable, but she couldn't actually cope with the baby. I had a good report from her college (NNEB) that she could. But she couldn't. As soon as she came she got into the clique of nannies round here and she found it irritated her when she wanted to go to sleep. And she didn't let her go to sleep. She took her to these social things. And she ended up crying. And it got into a vicious circle of crying, which I only discovered when the health visitor came round and told me that my baby was screaming down the area. . . . Really this particular girl was very immature beneath quite a brash and confident exterior. It took me a long time to see how immature she was. Also, she didn't know about basic nutrition. She didn't know what protein was for example. And I was staggered because Alice moved onto solids fairly fast. And I used to ask her what she'd eaten, so that I wouldn't feed her the same. And I was saying to her to give her some protein, vitamins and carbohydrate. And she didn't know what I was on about. I really got quite worried. I think they must have done nutrition but obviously she hadn't soaked any of it in.

We have here, then, a situation in which Janet Jones had no trust in the nanny to whom she had entrusted Alice's care, to the extent of being fundamentally concerned about her knowledge-base (and therefore competence). Even more important, however are the implicit criticisms of the health visitor. A crying baby is equated with an unhappy, inadequately cared for baby and – even though Janet Jones was not the daytime child carer – it was she who felt the failure. In failing to provide a good carer for Alice, she considered herself to have failed in her duty as a mother. Not surprisingly in such circumstances, the nanny was dismissed. By way of a contrast, these are Janet Jones's comments on the experience with the household's current nanny:

> Well initially I discussed with her the type of activities, and we discussed it a lot. 'Would it be a good idea if I took her to soft play?' 'What's soft play?' I didn't really know so I went down and had a look at it. 'Yes, I think that's a good idea.' And originally I wasn't very keen on all this socialising, but I realised the benefits to Alice, as well as to Carole. And I realised that I wouldn't want to be stuck in a house. . . . And then, as things have gone on, I've loosened up. I trust her. Originally I used to say 'I want to know where she is every day and I don't ever want to be in a position where I don't know where you are.' That was how we started. So she used to say, 'This is where I'm going today and this is the phone number.' But now I trust her. She exudes confidence and professionalism. So quite quickly I trusted her.

Between them, Janet Jones and Carole have negotiated the type of activities which now constitute Alice's days and weeks. Both nanny and employers (particularly Janet Jones) have a clear understanding and respect for each other's concerns and wishes. Trust is a central structuring principle of their relationship.

The Collinses

Andrew Collins is an investment banker and Pauline Collins is a medical researcher. They have two children, one of whom is just school-age. The household also includes a live-in nanny, Julia. As with almost all of the partnerships interviewed, the Collins's chief concern in employing a nanny was that the children were well cared for, emotionally secure and socially stimulated. As Pauline Collins explains, Julia is an excellent carer, but not so imaginative on the educational side:

INTERVIEWER: Did you have a clear idea too of what sorts of activities you wanted the children to do in a day?

PC: I organise – Chris has quite a few friends and so does Hannah. And most days after school he plays with somebody. And then, if they ask to be entertained, Julia does that. But it's usually my suggestion as to what she does with them if I know they've got nothing to do.

INTERVIEWER: So you're very much in control of their more educational activities?

PC: Yes. Although I'm not so concerned now as I was before Hannah went to nursery. Then I worried that she wasn't having enough educational stimulation. Because, although Julia is an exceedingly good carer – and the children *adore* her (her emphasis) – she's not very artistic. Whereas my previous nanny was amazingly artistic, but she wasn't so good a carer. She got more frustrated and shouted more at them. But Julia never shouts. She never raises her voice. She's so gentle and patient with them.

Later, Pauline Collins states that she wishes Julia 'to fulfil a mothering role for my children more than anything else'. The situation which exists, then, is one where Pauline Collins trusts Julia totally with the everyday love and care of her two children, and leaves this side of things completely in her hands. In terms of other activities however, she is much more directive and interventionist. In effect, Julia puts into practice Pauline Collins's suggestions. The situation here therefore is somewhat different to that with the Joneses. Although trust and consensus are as central to the Collins's arrangements as they are to the Joneses, Julia is far more of a direct substitute for Pauline Collins than Carole is for Janet Jones.

These three households highlight the importance of trust, consensus and negotiating frameworks to nanny employment within contemporary Britain, as well as their centrality to its day-to-day reproduction within individual households. And, as the case of the Jones household illustrates, it is the absence of these structuring principles which brings about conflicts and difficulties between nannies and their employers. But it is what is happening beyond the day-to-day level that is more important here. To be sure, the presence of trust, consensus and negotiating frameworks within nanny employment means that nannies and parents are in broad agreement over what is being done with children in the parents' absence. However, and critically, these factors also mean that the nanny becomes simultaneously both a childcare professional and a substitute carer. *In being either 'in tune with' parents' wishes (the Joneses) or doing what parents request (the Maxwells and Collinses), the childcare professional fuses with, and is subsumed within, employers' notions of care.* In what follows we examine why and how this fusion occurs.

The argument which we put forward here is that the fusion and elision between the childcare professional and the mother substitute reflects the mediation of the nanny form of waged domestic labour by the ideologies of mothering and motherhood. To elaborate, and to take the employer side of things first, we consider that it is the ideological construction of childcare as something which should be home-based, child-centred and performed by the child's natural mother which is the dominant force behind the resurgence in demand for the nanny form of childcare provision in Britain through the 1980s. The nanny – as a form of care based in the parental home and focusing, for the most part, on the care of one household's children – satisfies the first two of these ideological constructs. In that, by and large, most nannies have a professional qualification, it is also a qualified form of care. Using qualified labour in this capacity is readily understandable: quite simply parents wanted the 'best' form of care for their child(ren). However, we were left in no doubt that, in substituting this particular form of care, parents (and specifically mothers) were able to argue that children experienced a better form of care than that which could be provided by the housewife/mother. The Keiths, Lyles and Maxwells provide typical examples of such arguments.

The Keiths

MK: To be honest, she's probably better off educationally with some-one like Lizzie than if I was looking after her. *Because I'd be doing other things* [our emphasis]. I wouldn't be devoting all my time to her. Lizzie reads to her for hours and hours. And at twenty months she knew the difference between red, yellow, green and blue. She has had a lot of attention. And she is very happy. She's very gregarious and she'll go to anybody. She's not clingy to mummy.

The Lyles

SL: Once Diana had settled in it was nice to go to work and come home and have two happy children who had had a wonderful time all day, been well looked after and had lots more fun than they would have had if I had been at home all day, *as I wouldn't have been here only to look after them* [our emphasis]. I would have had other jobs. But Diana is there just to look after them, entertain them and give them a good time.

The Maxwells

CM: With our first nanny we had a really good arrangement here. We had two children who really enjoyed and got a lot of stimulus from each other and a person who spent her undivided attention on the pair of them. And when I looked at it I thought actually they were getting more attention and more stimulus than they would have done if I had been at home all day. *There would have been other things to do* [our emphasis]. And it just worked out really well. I think the children have benefited enormously from having another adult in their life who has taken on that caring role.

Given the equation of housewife and mother, the employment of a nanny – and consequent separation of childcare from other domestic tasks – is considered to result in a better form of care to that which a 'normal' mother at home with young children would provide. Professional (as opposed to untrained) and undivided in terms of the attention accorded to the child(ren), the care of the nanny is considered to replicate, yet surpass, the care which the housewife/mother can provide. Here then, immediately, in the minds of employers, is the conjoined childcare professional/mother substitute – the professional who is more than the mother . . . or is she?

When we look at the qualities beyond training which our interviewed households looked for in employing a nanny, conventional constructions of mothering were well to the fore. Here, for example, are Richard and Elaine Evans discussing their first nanny:

EE: She just loved children. *She was just the archetypal mother figure wasn't she* [our emphasis]?

RE: *She was as wide as she was tall – fifty-two, double D!!* And she had this little child on her knee who just went zonk. She was happy and content and that was it. Very placid, very calm, very relaxed. . . . She'd got Lauren on her knee and then all of a sudden Lauren made a noise as if she was about to throw up. But Sally – there was no reaction. And we looked at each other, and there was no question. That was it. If she could be that unflappable.

For Richard and Elaine Evans, as for many of the households we interviewed (see, for example, Pauline Collins on Julia above), it was Sally's physical and temperamental conformation with the traditional 'mother figure' – rather than her qualifications – which led to her eventual employment. Interestingly too, the conformation of the nanny with the 'archetypal mother figure' was something which many interviewed households contrasted with the mother's disinclination for the role. Here, for instance, are Stephanie Lyle and Corinne Maxwell on this theme:

SL: She has got much more patience than I have. Doing puzzles and reading, I'd be fidgeting and wanting to get on with something else around the house.

181

CM: I happen to believe that they are better cared for by somebody else other than myself because I am not interested in childcare. I do not enjoy trips to the park. I am very impatient with them if I have them for too long or all of the time.[3]

It is important to be clear about what precisely is going on here. On the one hand, as the above two comments indicate, the households we interviewed did not accept at a personal level that mothering is a 'natural' female characteristic. As households comprising 'working mothers' this is not surprising. Indeed, the entire form of the dual-career middle-class household presumes a rejection at the individual level of the dominant ideology of childcare, namely that the best form of care is that provided by a child's natural mother. At the same time, these households' comments on many of the nannies whom they employed suggest that childcare is something which they saw 'other' women as being *naturally* suited to – temperamentally and/or physically – as well as trained for. Given this, the employment of a nanny by middle-class households in contemporary Britain has to be seen to involve not simply the physical substitution of a predominantly qualified form of waged domestic labour for the unqualified, unwaged labour of a mother, but an ideological substitution too. Women who are deemed to have mothering qualities are taking the place of those who – for whatever reasons – represented themselves as lacking in this department. *In such circumstances the nanny is a mother substitute.*

The same elision between the childcare professional and the mother substitute within nanny employment was felt by nannies, who expressed precisely the same ideological constructs. Nannies, particularly those with the NNEB qualification, are encouraged (whilst training) to see themselves as childcare professionals, and the same view was promoted to us by NNEB course tutors.[4] Indeed, the latter were keen to draw attention to the way in which the NNEB course is designed to produce individuals with a sound knowledge of childcare, social and psychological development, health and education. Whilst this professional self-image goes unchallenged in the pre-qualifying period, both the reality of the labour market and nannying itself provide serious challenges to this fragile identity. In the first case, it is the ability of young women without the NNEB qualification to gain jobs as nannies and to command the equivalent (or more) in wages which erodes the self-image.[5] However, what happens within nanny employment itself may be even more important in challenging professional identity. Located as it is within the parental home, rather than in either a non-domestic environment (day care) or in their own home (childminding), the nanny is *the* form of childcare employment which most clearly expresses mother substitution. As we now show, the daily structure and organisation of nannying is one of the most important means through which this form of substitution is made visible to nannies.

Without exception, in all of the extracts of 'typical days' quoted previously, the basic structure of the nanny day revolves around taking over from the mother at the start of each day and handing over to her at the end. Although this was not always the case, in that in certain of our interviewed households (a minority) the male partner took a more active role, such was certainly the dominant pattern. This itself is symptomatic of the construction of nanny employment as between women. And, in almost all the cases we encountered, it was also the female partner who took responsibility for the day-to-day organisation of the nanny, and the female partner whom the nanny saw as her main employer. The Joneses and Maxwells are typical in this respect.

The Joneses

INTERVIEWER: And whose day-to-day responsibility is she?

JJ: She's mine, yes.

INTERVIEWER: And would she regard you as her 'boss'?

JJ: Yes. She'd regard first me and then Peter. She would come to me for most decisions.

INTERVIEWER: And are you the one who pays her as well?

JJ: Yes. It's a relationship between us two. Yes.

The Maxwells

INTERVIEWER: On a day-to-day basis, whose responsibility is the nanny?

CM: Probably me. She talks more to me.

PM: I pay her and she tells her what to do. That's life.

INTERVIEWER: Who does she talk to on a day-to-day basis about what the children have done and so on?

CM: Me, probably because I ask.

INTERVIEWER: Is that because you're here and Paul isn't?

CM: No. When the nanny is in the house and Paul comes in he usually goes up to his study and I'll have a coffee and a chat with the nanny and find out what's been going on.

Some important reasons underlie such day-to-day patterns of organisation. Significant amongst these are maternity leave itself, and the onus which this places on the female partner within a dual-career household to search for, and subsequently recruit, a nanny. But, beyond this – once more – are the all-pervasive ideological constructs of mothering and motherhood.[6] We will take each of these in turn.

Search and recruitment strategies

Interviews with both nanny agencies and employers showed that typically it is the female partner within a dual-career household who makes the initial contact with either a nanny agency or potential nanny employees. The woman is assisted in this task, in the first instance, by health visitors and other childbirth professionals, as well as by other organisations with which she may be involved, for example, the National Childbirth Trust (NCT) and its offshoot Working Mothers' groups. Following this initial period, and the drawing up of a list of potential employees (or agencies who will provide such a list), both partners are usually involved in the interview process. For example, Stephanie Lyle – whilst on maternity leave – contacted a number of nanny agencies and arranged visits to the family home by a shortlist of nannies whom she considered to be potentially employable. Having made her preliminary selection, James Lyle was brought in – in a managerial capacity – to oversee the final employment decision. Diana, their nanny, describes this situation:

> I went to X Agency on the Tuesday morning and she phoned up the lady straight away. . . . And she said, 'Well why don't you come round at six o'clock?' So I went round that evening. Well her husband wasn't home from work then so I saw Stephanie on the Tuesday and she said, 'Oh well, we've seen everyone now, I'll get in touch.' Well I got a phone call the next day saying, 'Could you come and meet my husband today? I've made up my mind but obviously he's seen all the other nannies and wants to see you as well.' So I went round there and spoke to him. They were both there then. And then Thursday she phoned up to say, 'Would you like the job?'

In some cases, however, the male partner is either not even involved in the interview of potential nannies, or – at best – is used in a rather more marginal role; for example, either to show interviewees round the house or in a vetting capacity. This seemed to be particularly so in the case of nanny-shares. Here Tracey – the nanny to the Newells and Orwins – describes her interview:

INTERVIEWER: So, getting back to when you started the job. Were you interviewed?

T: Well I went round to Alex's house and met both the mothers and they asked me all sorts of questions about what I had done in my other job and what training I had. And we had coffee and biscuits. It was just like a chat really. And they gave me the job there and then and I started on the Monday.

The above suggest some of the processes through which nanny employment comes to be constructed as between women. However, although search and recruitment strategies enable nanny employment to be constructed on mother substitute lines, the case of the Evanses illustrates that there is no inevitability to this.

The Evans's search and recruitment strategy was again one governed by maternity leave arrangements, with Elaine Evans approaching agencies and arranging interviews. Richard Evans played an overseeing/vetting role in all this but since then he has played a much more active part in day-to-day arrangements than most of the male partners interviewed. This is revealed in the following extract:

EE: I think the girls themselves stand more in awe of me than they do of Richard. Richard can say a lot more to them than I can. If I was to pass some comment like – I don't know – that they didn't want to hear, they'd take it a lot harder from me than from Richard.

RE: I tend to be the one who covers most of the disciplinary matters, partly because I take them home at night and so I have half an hour or whatever in the car on our own, and you've got to talk about something. So I was the one who drew the short straw to raise the points about behaviour, performance, discipline. But Elaine agrees what should be said, and when and how. . . .

INTERVIEWER: So if you asked any of the three who they regarded as their 'boss'?

EE: They would probably regard me.

RE: I think they'd say you, but I bet they'd hesitate. If you asked the question, 'Who's your boss?', they'd say Elaine, because she pays them But if you said, 'Who do you work for?', they'd say Richard and Elaine Evans.

Given that there is no necessary connection between search and recruitment and the day-to-day organisation of the nanny, it follows that something else is shaping of nanny employment in terms of mother substitution. It is hardly surprising that this something else should turn out to be the ideologies of motherhood and mothering.

The ideology of motherhood, mother substitution and 'the working mother'

One of the most fundamental developments within post-war Britain has been women's increasing labour force participation. There have been many conceptualisations of this. However, for our purposes it is the construct of 'the working mother' which is of most importance. Centrally related as this is to the dual-career household form, at its simplest the term does no more than refer to observable circumstances, i.e., to the mother who works. But beyond this it brings together two terms which – given the patriarchal form of gender ideology in contemporary Britain and the ideology of childcare in particular – are potentially contradictory, as well as morally and emotionally loaded. This construct of the working mother plays a central role in constructing the day-to-day reality of nanny employment on the lines of mother substitution.

As some of the previous employers' comments in this chapter suggest, a nanny is considered to offer quality childcare – a trained service which closely approximates care by the natural mother. As such, the employment of a nanny offers one of the easiest ways (ideologically, if not financially) through which dual-career households can negotiate both the construct of the working mother and the slur on working mothers. It allows women in dual-career households to see themselves as both 'the good mother' and the successful working woman. However, the ideological context within which this negotiation occurs also generates considerable pressure on the day-to-day form of nanny employment. Indeed, in many circumstances the very construct of the working mother helps to ensure not only that nanny employment is structured as between women, but that it is conceptualised in terms of mother substitution (rather than parental substitution). *To be a success as a working mother means that the onus is placed on women (rather than parents) to organise the replacement form of childcare.* As the following examples show, such pressures were felt by many of the women we interviewed, and led nanny employment to be constructed as between women, even in circumstances where it was perfectly possible for male partners to play a key role.

The case of the Collinses is particularly instructive. Describing her return to her job after maternity leave, Pauline Collins recalled how she had felt that employing a nanny was something which *enabled her to work*. In turn, this enablement led to the organisation of the nanny as her responsibility – to the extent that it is *her salary* which pays for the nanny. Other couples made similar comments, and in many circumstances it was the female partner's salary which was being used to meet the childcare bill. In some circumstances, such as with the Evanses, this appeared to be based on practical considerations.

EE: It's my job to pay her. I sign all the cheques. That's my job.

INTERVIEWER: How did you decide that that was your job?

EE: I don't know. It comes out of my pay anyway.

RE: All my pay goes on standing orders and everything else ... also it may be as simple as Elaine – when she was on maternity leave – had a bit more time to study the forms that go with it.

However, in other cases the source of nanny pay was more symbolic. Maria Brown, for example, remarked:

It works out at around half of my take-home pay. As long as I can take home more than half then its worth my while working, though of course Ken is working as well.

Here, much as with the Collinses, the nanny is seen to enable the female partner (and *not* the male) to work, and the arrangements regarding pay echo this. Similar thinking is also conveyed by the Harrises:

AH: I think the nanny is my role and my responsibility.
INTERVIEWER: Is she your financial responsibility as well?
AH: Yes. Nannies always have been. Ian's salary deals with other things. The nannies come out of my salary. I suppose it's because childcare is called women's work. And I suppose the nanny was to help me so I could work, not really for both of us.

Normative expectations regarding childcare and women's employment then lead nanny employment within dual-career households to be conceptualised predominantly in relation to the labour market partici-pation of the female partner, and not in relation to both partners. Furthermore, enablement carries with it connotations of obligation – the nanny becomes primarily the female partner's responsibility, a situation which is cemented in many cases by financial exchange between the female partner and the nanny. The concept of the working mother then, not only provides the rationale for nanny employment, it helps female partners to structure this employment as their responsibility and as a direct substitute for their own labour. But, as we now show, certain actions of male partners within these households reinforce this construction.

Clive and Barbara Roberts are both university lecturers. This form of employment permits Clive Roberts to work from home a good deal of the time. Here this partnership discusses their differing relationship to nanny employment.

CR: I feel strongly. I feel resentful if they come and see me during nanny hours [he is referring to the children here]. I have got my work to do and if I am in my study and they come and break into the study I am peeved because the nanny should be looking after them. They should not be breaking into the study and disturbing me.
BR: [to Clive] I think you feel that more than I do, partly because I always like to talk things over more at the end of the day with

Rachel, as to how things have gone and what has been happening in the day.

CR: Whereas I very much consider that from eight-fifteen to five-thirty I have nothing to do with the children . . . I mean if he asks me or if either of them ask me if they can do something between eight-fifteen and five-thirty I always say 'Ask Rachel'. I do not consider it my function – unlike Barbara – who has more of an overlap at the beginning and end of the day. I just stop at eight fifteen and switch on again at five-thirty in respect of being a father.

Interestingly, Rachel inadvertently confirmed this situation in the course of our interview with her:

INTERVIEWER: What about overtime. Do you ever end up staying a little bit longer?

R: I've never done it. I think once they've been five minutes late. He just doesn't do much work anyway. He goes in at the last minute just before he has a lecture and then, once he's finished lecturing, he comes home again.

INTERVIEWER: What happens for example if he comes in at four?

R: He goes upstairs and watches telly. And he'll come down at twenty-nine minutes past five.

These circumstances are by no means atypical. As well as the few who worked in the main from home, in several of the households we interviewed the male partner was as likely to come home early as the female partner. In none of these cases, however, did his return signal the end of the working day for the nanny. By contrast, the early return of the female partner usually resulted in the nanny's going home early. Such practices demonstrate the construction of the nanny as a mother substitute, rather than as a substitute for parental care. Moreover they also accord with the minimal engagement which we encountered between fathers and nannies. Indeed, although there were the few exceptions, for the most part fathers' involvements were confined to disciplinary matters, dealing with cars, buggies and pushchairs when they went wrong and acting as a mouthpiece for the female partner in her absence.

The foregoing shows how nanny employment within dual-career middle class households assumes the mother substitute form and begins to suggest how nannies themselves come to experience this form of work as being about mother substitution. But, rather than being passive participants in this process, nannies themselves engage with some of the central structuring principles of motherhood and mothering, thereby undermining the already fragile self-image of the childcare professional and reinforcing mother substitute notions.

Four instances reveal ways in which nannies do this.

Carole (the Jones's nanny)

INTERVIEWER: . . . and this job, does it involve housework?

C: Well they have a cleaner in once a week *but I would normally hoover through once a day downstairs.* I don't do any dusting. *I do the washing* but Janet does her own ironing. *And I do a bit of shopping now and again* [our emphasis].

The Keiths (with Lizzie)

MK: She's 23 going on 50. She's very, very sensible. Far more sensible than I am. As an example, this morning she said to me, 'You need some stuff for the freezer. If you give me the money I'll go to the shops and get it. And then I'll write you a list of the things that you need to get at the weekend when you go to Tesco.'

Karen

Well I've never actually been told 'you've got to iron their clothes', it's just with May-lou starting school and some mornings she's wanted a dress and it's never been ironed. . . . Cheryl will say, 'I'm just ironing this' but I feel it should be me and I'll say, 'Ah no, I'll iron it.' And I thought to myself, 'This is getting a bit silly.' So, off my own bat I started ironing. . . . I tidy up. I don't dust. I don't hoover – unless there's something spilt. But I pick clothes up, towels – stuff like that – and I'll put the papers away. It's not that I've been asked to do it. It's just the way I like to work.

Marie

INTERVIEWER: Do you actually do any housework?

M: It doesn't. In my contract it says all I have to do is clear up after meals and make sure the playroom is tidy. But I do a lot more than that because, because basically I care. I care a lot about them – the family. And I know that they don't get enough time with each other. I do things like if there's washing in the washing basket I'll put it through and hang it up, or empty the dishwasher. But then sometimes I'll do the whole house and you know, vacuum and clean the floor.

As these extracts show, nannies often do more in the way of domestic work than simply those reproductive tasks pertaining exclusively to childcare. Furthermore, in all four of these cases these activities were *not* part of their job description, but were proffered voluntarily – a situation which contrasts markedly with those circumstances in which households required general domestic work of their nannies.[7] Part of the explanation for such practices relates to something which Marie refers to, namely caring for and about a particular household, but the behaviour also reveals that nannies identify with the role of mother substitute. In performing domestic chores, nannies are doing no more than what a 'normal' mother at home with children would be doing. Though the occupation of the nanny rests on the subdivision of domestic tasks between those pertaining to children and those relating to other household members and the home, in practice this subdivision (which many parents wished to preserve) is being eroded by the voluntary practices of nannies themselves. Seeing themselves as mother substitutes enables nannies to fuse the practices of mother with those of the housewife. Regardless of why this happens within individual instances, there is little doubt that – in so doing – the nanny comes to see herself as *both* a mother substitute and a childcare professional.

'SOMEHOW IT'S MORE THAN AN EMPLOYER AND AN EMPLOYEE ...': THE SOCIAL RELATIONS OF NANNY EMPLOYMENT

> She is much more than a nanny. It's a terrible cliché, but she actually is part of the family.

Although, by definition, a waged form of domestic labour, the social relations of nanny employment are characterised by an elision as fundamental as that between the nanny as childcare professional and as mother substitute. This is captured in the above quotation: the social relations of nanny employment are constructed by and shaped through wage and false kinship relations. In this section we discuss each of these relational constructs in turn, before moving on to consider the implications of false kinship relations for both nannies and their employing households.

As waged labour, nanny employment is shaped at one level by relations between employers and employees. Appendices 6.1 and 6.2 establish the material basis of this relationship. Although there were differences between our two study areas in pay levels, nanny employment is characterised by low-paid, for the most part contractual, work. It also involves certain of the employment conditions associated with other forms of waged work (notably paid holidays) as well as – in some instances – a range of 'perks'. These arrangements suggest a fairly straightforward, contractual form of low-paid work. However, when we analyse the exchange relation between nannies and their employers, the

190

distinctive nature of the social relations of nanny employment starts to become manifest. Although a number of employing households paid their nannies formally (i.e., the wage received went through formal channels and included tax and National Insurance contributions), a sizeable number did not. In these circumstances the nanny received a standard 'declared' payment, together with cash 'top-ups'. The employers' rationale for so doing was, of course, to avoid what they saw as 'double taxation'. Having paid tax and National Insurance contributions out of their own salaries, they did not wish to do the same again in connection with nanny employment. However, of particular interest was the way this situation was presented to nannies. Rather than explaining things in terms of their personal finances, employers tended to present this arrangement as a favour beneficial to the nanny – typically in the form, 'Wouldn't you rather have the extra "cash in hand" than lose this through deductions?' The employer is seen to be 'doing a deal' with the nanny; one which in the short term appears to her advantage, but one which – long term – clearly is not. Such arrangements indicate the way in which – even at the material level – nanny employment can be permeated by notions of 'helping out' and 'favours'. Visible from a very early stage in this research, 'doing deals' was a clear pointer to the importance which false kinship relations were to assume in the later stages of our analysis.

As we proceeded beyond the material level to examine the social practices of nanny employment, the elision between the social relations of waged work and those of false kinship became increasingly apparent. Indeed, as we now show, the day-to-day social relations of this particular form of waged domestic labour fuse both sets of relations. The Harrises provide one of the clearest instances of this double construction.

Anne Harris supervises nanny employment in negotiation with the nanny – Frankie. A central facet of their arrangement is the keeping of a diary relating to Josh's activities. Frankie started keeping this diary when Josh was a small baby. At that time the diary focused on things like sleep patterns and feeds, and later on on the foods eaten. Now it is still used to record meals, but has been expanded to include the range of activities in which Josh participates each day – swimming, soft play, etc. This diary is Anne Harris's means of keeping an eye on Josh's everyday life but it also enables her to keep tabs on what Frankie is doing with him. In this respect then, Anne Harris is clearly playing the role of the employer.

In the course of our discussions with the Harrises it became clear that the above supervisory device was indicative of a relationship structured on employer–employee lines. This construction is well to the fore in the following extract:

191

I think the relationship with her has been fairly good but there has been a tendency to be somewhat inflexible – particularly at the end of the day, which is a problem. I'm self-employed and so is my husband, and there are times when I have to stay late and he has to too, and we have to say, 'Frankie, there's a problem. Can you hang on?' Which obviously you don't want to abuse. I have arguments about this with my husband too. He says that she's there for our convenience. 'If we weren't working in these jobs then we wouldn't need her. You can't let her times dominate your life and run your business'. . . . Part of the problem now though is that her fiancé works at X now . . . and he actually works shifts. So now they have this little window in the day and I'm very conscious that I don't want to encroach on that. It's between six and seven. But, if there was a little bit of flexibility there then it would be better. So this week in fact we've – not exactly had words – but reminded her that we do have problems at these times.

Confronted by a problem, it is ultimately the power relation embedded in the employer–employee relation which enables Anne Harris to act in the manner she describes. Nevertheless, she is far more concerned than her husband about respecting Frankie's rights – not just as an employee, but in relation to her social life as well. These comments provide the first signs of the double construction present within this individual instance of the social relations of nanny employment. Shortly afterwards Anne Harris expanded on these ideas:

It's difficult. We are obviously employing her to do a job but we want to be friendly towards her, and we want Josh to be happy with her as well. She invited us to her 21st birthday party and that sort of thing. . . . I think the underlying part of the friendship is that she and Josh get on so very well. And therefore we're friends because she's friends with Josh. Which I'm reasonably happy with. If I'd met Frankie anywhere else we'd have had absolutely nothing in common. We would have thought, 'She's a nice girl'. And she'd probably have thought, 'She's reasonably OK'. And it would never have gone any further.

From this we get a clear picture of the 'more than an employee' overtones which figure so strongly in the social relations of this form of waged domestic labour. For the Harrises, the friendship between Josh and Frankie acts to complicate the employer–employee relation outlined initially. In effect, the bond between Frankie and Josh extends to bind Ian and Anne Harris in a set of false kinship relations – relations which exist alongside and – for the most part – override the bottom line employer–employee position.

A similar double construction is revealed by the Frazers. For Ruth Frazer, as for Anne Harris, 'The bottom line is that I'm the employer, she's the employee. I can tell Debbie to do something and it doesn't matter how nicely I say it, I expect her to do it. She's never said, "I don't let the children do this. Can we please be consistent. Would you please not do that".' But Ruth Frazer also acknowledged the existence of the false kinship counter-construct:

> It's a very personal thing. It's not as cut and dried as most jobs because she's coming into my life much more than anyone in the workplace would be. That complicates things. . . . Debbie comes in and cries because a bloke's done this and that to her and you're really personally involved there.

Repeatedly in our interviews we encountered this elision in the social relations of nanny employment. Although they are employers and employees, employing households and nannies are clearly bound by relations which extend beyond those of wage labour. In what follows we explore these false kinship relations in more depth.

The relations which the above two instances suggest are clearly *not* those of kin relations; they are not constructed between individuals related through blood and/or marriage. However, if we move beyond the level of who is related to whom, to examine the practices of nanny employment, it is evident that this form of waged domestic labour exhibits certain facets of the type of support conventionally associated with kin networks (for a discussion of these, see Finch, 1989). Definite patterns of familial incorporation are involved in nanny employment. Moreover, the employing household is usually involved in at least one, and sometimes two, of Finch's forms of kin support, namely emotional and moral support and accommodation.

To take the familial incorporation issue first. During our fieldwork we encountered a number of descriptions by employers which emphasised the kin relation theme. Thus, nannies were represented to us as 'a sister' to the female partner in the household; as 'a big sister' to the children in their charge and even – on one occasion – as 'a wife' to the female partner. Here, for example, is Marianne Keith again:

INTERVIEWER: How would you describe your relationship to Lizzie?
MK: I sometimes describe it as she's the next best thing as a wife to me. She is brilliant. . . . She's really like part of the family now. I actually confide in her – things I wouldn't necessarily tell anyone else. She knows so much about my private life that even my close girl friends here don't know about.

Others – like the Lyles below – represented things in terms of both a very close friendship and familial incorporation.

SL: We like to feel that the nanny is part of the family, not an employee. She's part of the family.

JL: We don't actually think of it as employing her somehow. It obviously is and it's all financially straight and she works set hours but we don't think of it in terms of an employer–employee relationship.

SL: No, with Diana it's more like a friend coming in to look after the children now. Obviously initially it was much more formal because we didn't know her then. But as time has gone on . . .

JL: So that's how we perceive it. Hopefully Diana feels fairly relaxed about it as well. If it was too formal it wouldn't have fitted us. We don't think of her as being staff. She's there helping out and we pay her for that.

However, in the case of the Maxwells, the ties were – if anything – even closer. Here Corinne Maxwell describes her feelings towards Janet, the household's live-in nanny:

> She is more of a sister to the children and to me. She is a lifelong family friend and if I dropped dead tomorrow the children would go and live with her.

We did not ask the nanny concerned what she felt about this proposal! Nevertheless, Janet is currently living with the Maxwells even though they are no longer her main employers.

So much for the employer perspective. Nannies viewed the situation somewhat differently. Few used the label 'friend' voluntarily to describe their employers, and even fewer saw things in the guise of a sister to either the female partner or the children. Instead, they were more likely to represent their employers, firstly, as the people for whom they worked, and, secondly, as people with whom they were friendly. In the majority of cases then it was the employer–employee relation which was being described as 'friendly' by nannies. Here, for example, is Karen again:

> They don't like – they never give you orders or anything like that. Like I remember when I had my second interview to get the job, Cheryl – one of the questions she asked was, 'What kind of relationship would you imagine we'd have?' Or something like that. And I was saying things, and she says, 'It should be more like a sister relationship.' But we get on well compared to some of the people I talk to.

There are important subtleties which differentiate Karen's comments from employers' representations of relations with nannies. 'Friendliness' here refers to 'getting on well' i.e., to harmonious, easy-going relations with employers. *It is not being used to refer to a relationship beyond*

194

employment. Moreover, the contrast with the employer view – although largely implicit – is there to see. Although Karen's female employer represents her relationship to Karen in terms of pseudo-sisterhood, the relationship is not the same in reverse. We encountered the same differential representation time and time again.

As might be anticipated, social relations were visible in the practices of nanny employment. Thus, and almost without exception, nannies were present as honorary family members at children's birthday parties; and we even encountered one instance of a nanny being given away at her wedding by the male partner in the employing household. We also found clear examples of employing households (and particularly female partners) acting in an advisory/supportive capacity towards their nanny.

More common, particularly in academic households, was the encouragement of educational advancement. Janice's employers, for instance, are trying to encourage her to take a couple of mornings off each week to study.

All this points to the incorporation of the nanny within the familial activities and practices of employing households. However, at the same time it became apparent that whilst in one sense 'insiders', nannies were also 'outsiders' within employing households. Some of the clearest expressions of this were evident when kin relatives – particularly grandparents – came to stay, in which case the nanny was usually given 'a holiday'. Although represented by employing households as part of the family, the status is fragile. Nannies constituted substitute kin in periods when kin relatives were unavailable, yet lost this kin status for the periods when the latter became available.

At the same time as some of the social practices of nanny employment confirmed the false kinship representations of employers, other social practices revealed the family of the nanny to be incorporated too in a complex web of false kinship relations. This was particularly so with respect to the children involved. Here, for example, is Karen again:

K: I take Danny and Katherine – Danny and Katherine get on well with my family, and they class them as 'auntie' this and 'uncle' that.

I: So do you take them quite a lot to see your mum and dad?

K: Well, just two hours a week. On a Tuesday afternoon I take Danny. But me Mum classes them as her grandchildren sort of thing. They get Easter eggs and Christmas and birthday presents.

Much the same type of relations was described by Kay – who as well as meeting up with her mother on frequent occasions through the working week, had taken her charges to spend Boxing Day with her family. Interestingly, we did not encounter any instances of such practices of incorporation extending to taking employers.

In summary, then, whilst abstract representations of the social relations of nanny employment suggest a differential construction by nannies and their employers, social practices serve to counter this. Nanny employment in contemporary Britain is characterised by a complex web of false kinship relations which involve both the families of employers and employees, and which centre on the nanny and the child(ren) being cared for. Thus, the nanny is constructed as both insider and outsider by employing households, whilst charges become incorporated within nannies' own families, for the most part as pseudo-grandchildren. We consider the reasons behind this particular social construction at some length in the final section of this chapter. However, first it is important to examine the implications of false kinship relations for the day-to-day reality of nanny employment

As might be anticipated, the chief problems which false kinship relations posed for the nannies we interviewed were ones to do with employment conditions, and – not surprisingly given their earnings (see Appendix 6.1) – it was pay which constituted the major grievance of the nannies we encountered. Moreover, quite apart from the problems associated with the one-to-one negotiating position which characterises nanny employment, there seemed little question that false kinship relations compounded what was an already very difficult area for nannies. Indeed, the construction of the social relations of nanny employment on false kinship lines complicates things to the extent that to work as a nanny becomes, in effect, a labour of love. Diana's comments provide a classic instance of what we mean:

> When I had been here six months they were going through quite a bad patch. They'd re-mortgaged the house and whatever. And I knew it was coming up to six months, but I'd never had a pay review. . . . And I thought, I can't say anything, you know, I'm so happy that I would rather not mention it and stay, than sort of like open my big trap and then sort of have them think 'Oh God, she's only doing it for the money.' So I didn't say anything

Diana was given a £20-a-week increase by her employers, but despite this, after a year's employment, says:

> I don't know what is going to happen this time. I'm not going to say anything. I'm just going to wait for them to say it.

For her, like so many of the nannies we interviewed, the centrality of love and caring to the social relations of nanny employment mean that it is effectively taboo to raise the pay question. Indeed, for Diana to have done so – although she did not express it in these terms – would have meant to challenge the entire social relations on which this individual instance of waged domestic labour appears to be based.

Other financial aspects pertaining to employment conditions – for example, tax, national insurance and sick pay – were also difficult issues for nannies to approach employers about. And, once again, it was false kinship social relations which seemed to be getting in the way. Here is Tracey demonstrating her uncertainty in this area:

INTERVIEWER: Do your employers pay your tax and national insurance then?

T: I think so. Well they say they do, but in my last job they said they did and when it came to the end of the year and I left I got a big bill from the tax man. Anyway, these say they are doing. I suppose I ought to check really but we get on so well I don't want them to get annoyed by thinking that I don't trust them, because that's not really fair.

The type of trust displayed by Tracey towards her employers is characteristic of our fieldwork findings. Indeed, as our research proceeded, it became apparent that the presence of false kinship relations, the centrality of trust and 'getting on' to nanny employment and the consequent construction of nannying as a labour of love by nannies mean that this particular form of waged domestic labour is open to potential employer abuse. We encountered a number of instances of this amongst the nannies we interviewed, although – for the most part – these had occurred in previous jobs. In almost all cases they related to pay and national insurance contributions; although others occurred in relation to holiday entitlement. Here, for example, is Kelly, recounting what she discovered when she left her previous employment:

They gave us the P45 back. And I said, 'This is the same P45 as what I gave you.' And she said, 'Oh I know, but there wasn't any time to sort it out.' I was in two minds as to whether to tell anyone about it, but I thought it was best left. If I ever do get queried about what I did between such and such dates I will say.

In other circumstances it was nanny pay which was the occasion for employer abuse. Janice, for example, left her previous job when her then employers decided that they could no longer afford to pay her at a particular rate, and asked her to take a pay cut of £20 per week, as well as to move to cash-in-hand earnings. In presenting this package to her, they had clearly played on her feelings of love and loyalty to the children. With Hannah, however, the problem had been that on several occasions she had not even been paid.

I had problems with my first job. They couldn't afford to pay me. I'd get to my pay day and they'd say, 'Oh we're sorry you're going to have to wait. We can't afford to pay you.' I got very friendly with them and it seemed to me that in the end I was doing things for more of a favour to them because we were really good friends sort of thing.

For nannies, then, false kinship social relations are critical to enabling them to construct their work as a labour of love. But, once constructed thus, nanny employment becomes subtly differentiated from the world of 'real work'. False kinship relations, and the concept of a labour of love which these generate, effectively undermine and cloud the structural position of the nanny as *waged* domestic labour.

From the perspective of employing households, it was the same labour of love construct which could prove problematical, although it should be stressed that we consider these differences to be of a different order of magnitude to those confronting the nanny. For employers, there was little doubt that false kinship social relations proved, in general, highly beneficial. In effect, they enable the emergence of notions of love, trust and caring, which – in turn – we found to be expressed through the doing of 'extra work' by nannies. Normally routine housework, this 'extra work' also extended to include staying late and babysitting. However, false kinship relations *could* create areas of difficulty for employers. In our fieldwork it was primarily discipline, and the ways in which certain things were done – particularly with respect to children – which employers alluded to as areas complicated by false kinship relations. They did not see false kinship relations getting in the way of issues to do with employment conditions. However, and as befits their representation of the social relations of this form of waged domestic labour, employers were more than reluctant to discuss explicitly the difficulties brought about by false kinship relations, particularly in relation to their current nanny. Instead, they tended to refer in general terms to problems encountered by 'others', citing – for example – alleged instances of theft by nannies as illustrative of the way in which nannies could abuse employer trust and stressing how false kinship relations made dealing with such cases difficult and/or unpleasant. Although we acknowledge the potential for such problems within nanny employment, it needs to be emphasised that (at least in the course of our investigations) we found nothing remotely supportive of such allegations. Instead, the direct instances of problems cited by employers as clouded by false kinship relations proved to be rather more minor. Here for example, is the Quigley household:

SQ: I thought, 'She's my friend.' So there were things that I tolerated that another person wouldn't. Little things like chopping on work surfaces and not on the chopping board, so leaving scratches. It really got me but I never said anything in the whole five years.

Although apparently referring to an insignificant issue, the comments of Sheila Quigley speak volumes about how false kinship relations can get in the way of being the employer. Further confirmation of this came from our interviews with nannies, particularly where they referred to differences between their own (professional) views on child development and those of their employers. Louise, in the following extract, provides a classic illustration of the type of circumstance which we encountered on a few occasions.

> When I was potty training him, they would still put him in nappies, and when they take him out they still put him in nappies. He still goes to bed with a nappy on! And they still believe he needs a morning sleep. Sometimes he does. Sometimes he doesn't.

In this case there is a clear difference of opinion over what is being done when with a child, but rather than talk to the nanny about this (the reaction of a number of our employing households) this household – like others – responded by doing their own thing with the child outside nanny hours and by letting the nanny get on with her way of doing things at other times. In such instances there seemed little doubt that – for whatever reason – employing households were choosing not to exercise the employer role. In so doing, therefore, they remained in false kinship mode. Although the reluctance of our interviewed households to engage with these questions directly makes it impossible for us to discuss this issue conclusively, it seems that Sheila Quigley's comments offer some insight into what is going on here. The existence of non-confrontational relations in connection with different practices within some of our employing households may be indicative of the difficulties of being 'the employer' within a relation constructed for the most part on false kinship lines.

Clearly then, the existence of false kinship relations within nanny employment proved problematical for some households. In short, they made certain forms of intervention difficult for employers, although obviously – given the underlying employer–employee power relation – far from impossible. Moreover, these difficulties were ones felt more acutely by the female partners within a nanny-employing household. The comments from Anne Harris with which we began this section highlight this situation. As a consequence, it was by no means unusual to find instances where, although the day-to-day organisation of nanny employment was designated as 'between women', bigger issues – particularly those involving discipline – were demarcated as the responsibility of the more distant male partner. Indeed, we would argue that the distance of many of the male partners in our interviewed households from day-to-day nanny organisation meant that they were less embroiled within the web of false kinship relations and – as a consequence – more able to assume the employer role as and when required than their partners. Clare provides an excellent example of precisely this.

Clare recently married and, some months prior to her wedding, was trying to finalise the purchase of a house . . .

> We were getting this house and I needed just an hour off work. Obviously, you don't get a lot of notice when a solicitor says can you come in on such and such a day. And it was about two days' notice I could have given them. So I told them as soon as I knew. I rang her up at work and she babbled on about how she didn't think she could get the time off. I needed to be there for between nine and ten in the morning. So I said, 'Well, I could take him to me mother's', because he loves me mother. But no that wouldn't do. So she rang John to see if he could take the time off work. Then she rang me back and said, 'Look, John's going to see you when he gets in about it. We're both not amused by this. Two days isn't enough notice.' . . . So, of course, I was het up all day about it. And he came in and he said, 'I'm sorry but you can't have the time off. You're just going to have to let Craig go on his own.'

In this household, as in almost all of those we interviewed, the female partner deals with Clare on a day-to-day basis. However, as we see here, in an issue relating to employment conditions, the male partner is wheeled in to exercise the employer's veto. More generally, this instance points to the differential gendering of the social relations of nanny employment, a theme which we explore further later in the chapter.

A final area which needs highlighting with respect to employers and false kinship relations is one which pertains not to current circumstances but to the future. In a number of cases, nanny employing households made reference to uneasy feelings of responsibility towards their nannies. Here, for example, is Ruth Frazer on this theme:

> What I do feel is a more general concern. She's 23. She hasn't got a boyfriend. She can't find anybody she likes and most other 23-year-olds are either married or going steady or whatever. And she's sort of in limbo. She's brought up in X, and she's bored, bored, bored. She feels she doesn't actually belong anywhere. And I wonder if working for me has exacerbated this. That working for me has introduced her to another side of life.

Although initially Ruth Frazer professed not to feel responsible for Debbie's future, as she shows, she is in fact genuinely concerned about this and acknowledges her own part in making this more problematical for Debbie. Moreover, at the later stage in the interview, she remarked:

What I would like to happen is that she would become a child-minder in our house. She'd have a baby; she'd have another source of income and we'd not have to find somebody else for after-school care, which would be very difficult.

Although clearly a comment motivated by personal interest, the above also points to a general unease and concern about Debbie's future. It is not just that having Debbie around would solve the after-school care problem for Ruth Frazer. Rather, it is having Debbie around which matters too. Indeed, if *Debbie* wasn't important in this future scenario, then why not set up *any childminder* in their home to solve the problem of after-school childcare?

For employers then as well as employees, the reality of nanny employment in contemporary Britain is that it is heavily overlaid with the type of social relations more commonly associated with kin. In the final section of this chapter we examine why this is so.

EXPLAINING FALSE KINSHIP

In attempting to account for the significance of false kinship relations within nanny employment, our first thoughts turned to the geographical mobility of our north-east and south-east households. In common with the findings of Savage *et al.* (1992), most interviewed households were not living in their 'home' regions. As a consequence, kin relatives were unavailable to them on anything other than a short-term, temporary basis – typically around the period of childbirth. As the extent of false kinship relations within nanny employment became apparent we started to think in terms of the creation of substitute kin networks by our employer households. On reflection, we consider this 'first cut' into the explanation of false kinship to be deeply misleading. In what follows we argue why we believe this to be so, before examining the processes generative of false kinship relations.

Our case that the false kinship relations present within nanny employment in contemporary Britain do not represent an attempt to create substitute kin relations rests on the argument that the bonds of kin support do not rely on geographical proximity. As Finch (1989: 94) argues, 'At a commonplace level, geographical distance is bound to affect the kind of support which is possible between kin. . . . However, it would be a mistake to assume that there is a simple linear relationship between geographical distance and the level of support between kin.' Instead, the geographical proximity of kin relatives, or lack of it, simply mediates the forms of kin support available (and see too, Bell, 1968; Wallman, 1983). Although we did not interview households directly on the nature of their kin support networks, almost all made reference to parents and to sisters/brothers. We therefore consider it highly likely that kin support networks of the form discussed by Finch (1989) were present within these households, and that the partnerships which we interviewed were bound by feelings of obligation and responsibility to their relatives. Given this, there is no way in which the presence of false kinship relations in nanny employment can be interpreted as an attempt on the part of nanny employers to generate *replacement* kin relations. Indeed, it could be argued that the profound difficulties which many middle-class women working within full-time service class occupations face when confronted by the need to negotiate normative moral guidelines with respect to kin relatives (see for example Ungerson, 1987), mean that kin relations are sufficiently problematical in themselves, without generating substitutes. So why then do false kinship relations emerge in the context of nanny employment?

The explanation which we offer rests on some of our earlier arguments. Figure 6.1 summarises the false kinship relations discussed in the previous section. At the heart of this diagram are two key relations, those between the nanny and child(ren) and those between the nanny and the female partner within the dual-career middle-class household. As we now show, it is these two relations which are behind the production and reproduction of false kinship relations within nanny employment.

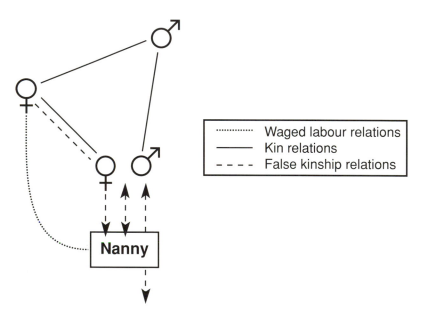

Figure 6.1 The social relations of nanny employment
Source: fieldwork interviews

To take the nanny–child(ren) relation first. There is little doubt that the relationship between the nanny and her charges is one of deep attachment. Thus, although the nannies we spoke to were well able to articulate in theory the dangers of over-attachment with their employers' children, the bond between the two was manifestly apparent, both in terms of what was said and what was done with children in our presence. Here, for example, are the comments of Rebecca, Louise, Sally and Marie on this theme. They are by no means unusual.

Rebecca: You get really attached to the children. It doesn't matter how hard you try not to, you can't. And even with Neil – I mean I didn't have Neil from a baby, I had him from one – even now I would hate to think of someone else looking after him. And that's what I was thinking when I was due to leave in September [for the B.Ed. course]. That was on my mind. And I used to think, 'Oh I wouldn't like to see him running up to some other nanny.' You know. 'Cos they are just like your own in a sense.

Louise: It's a weird job. You get so close to them. Like in my first year working here it was horrible when I took holidays. I just didn't want to take any holidays. You know what I mean? I'm all right now. Like I don't mind taking them but in the three weeks when I went away in the summer I really missed them.

Sally: I love Timmy . . . I adore him and he's like my own because I've done everything that a mum would do for her child, if you know what I mean.

Marie: A few months ago I had a big break-up in a serious relationship. And you know my first reaction was just to go, just to drop everything and leave. Go abroad, or anywhere, just to get away. But she was the one steady thing in my life. She was the one person who would always love me, no matter what I did. So I couldn't leave her.

As these extracts make clear, the attachment between a nanny and the children in her care constitutes one of the main reasons why one nanny will stay with a household over a period of years. At the same time, this bond was one of the reasons why others of those whom we interviewed had moved between nanny jobs. For example, Diana told us:

Although I've only been here just over a year, I don't think I want to stay that long because I think I'll just get too attached to them, you know. Whether they'll change as they get older is another story, but at the moment I'm very attached to them. . . . When you leave you know you think, 'Oh God, what am I going to do without her?'

Moving beyond the level of respondent comments, the ties of attachment felt by nannies we interviewed were ones which we felt reflected what Graham (1983: 18) has referred to as the way in which caring is 'given' to females as the defining characteristic of their self-identity. In reporting to us their visibly obvious pride in seeing and helping their charges develop, and in acknowledging the central part which their work played in this, nannies were not just responding to the individual circumstances of particular jobs, but expressing their identification with the construct of the female carer. This accords strongly with the type of emergent gender identities discussed in Chapter Five. But, perhaps even more interesting, is what this particular identification sets up. It is at this juncture that our previous arguments about mother substitution become important. Quite simply, the caring bond between the nanny and her charges is one which is *about specifics*. It is about what a nanny feels about a particular child. They are – as Rebecca and Sally say above – *'like your own'*. Given this, nannying has to be seen as an occupation permeated not just by notions of caring, but as *an occupation in which notions of caring can become equated with – indeed are highly likely to be equated with – substitute mothering*. In such circumstances it is easy to see how false kinship relations begin to emerge. In beginning to construct themselves as either the specific carers of specific 'others' or (particularly) as substitute mothers, nannies come to see their charges as needing, as being dependent on, themselves. As a consequence the relations which assume importance in nanny employment are those of emotional support and practical care – forms of support which traditionally characterise kin relations.

Referring once more to Figure 6.1, the second key relation within nanny employment is that between the female partner (mother) and the nanny. We now concentrate on showing how this relation itself is generative of false kinship relations. The key to what is going on here is, we believe, the construct of the working mother. As we demonstrated above, given the identification of the work of childcare in contemporary Britain with *mothering* (rather than with fathering and parenting), forms of childcare which go beyond normative guidelines are conceptualised in terms of mother substitution. As we saw, many of the female partners we interviewed internalised this construct. They felt that the employment of a nanny enabled them (and not their partners) to work and in many cases cemented their feelings by the financial organisation of nanny employment. It is important to be clear about the implications of this for nanny employment.

At one level, this form of enablement is not a sufficient explanation for the emergence of false kinship relations within nanny employment. However, what enablement points to is that having the nanny – *and specifically one/two particular nannies* – comes to be seen by women within dual-career middle-class households as the means through which they negotiate working motherhood. As a result a clear set of 'debts' appear to be established, some of which form the basis for the development of feelings of obligation by the female partner towards 'her' nanny. As we now show, these feelings of obligation work to produce social relations overlaid by false kinship.

As a form of waged labour, part of the debts involved in nanny employment are repaid through the wage relation. Others are emotional: we consider that they are the ones which generate the feelings of obligation which our respondent female partners either referred to or alluded to, and in turn, false kinship relations. It is here that the feelings of the female partner start to intersect with the relations between the nanny and the child(ren) outlined above. Indeed, we would go so far as to argue that it is precisely the bond of attachment between the latter which works to create feelings of obligation in female partners. Thus, in providing quality (as well as loving) childcare, the nanny enables the female partner (at least at the personal level) to negotiate the moral burden (and frequent guilt) heaped on the working mother. For many of the women to whom we spoke to therefore, it became necessary to repay this emotional debt. We consider the false kinship relations which characterised the representation of nanny employment by our female employers to be both a manifestation of this emotional debt, and their way of repaying it. Constructing nanny employment on these lines was their way of showing how much they cared about the work their nanny was performing.

If we are to account then for how and why the social relations of nanny employment within contemporary Britain come to be constructed in terms of both wage labour and false kinship but dominated by the latter, we have to recognise the way in which social relations within this occupation are produced by and through the ideology of childcare. As we have seen, the latter provides the framework which ensures that nanny employment assumes the form of mother substitution and hence becomes permeated by feelings of attachment and obligation *between women*. We consider false kinship relations between women to be an almost inevitable consequence. Moreover, this same construction lies behind the marginalisation of male partners (Figure 6.1) within the social practices of nanny employment. Conceptualised in terms of mother substitution (rather than parental substitution) nanny employment in dual-career households carries, for men, none of the connotations of emotional 'debt' and obligation discussed above. Instead, for them, the relation is simply one of waged labour – albeit located within their own home and albeit in relation to their children. Further, we feel that this differential gendering of the social relations of nanny employment lies behind the use of male partners in a disciplinary capacity by women. With a social relation grounded in wage labour, rather than one complicated by the ties of false kinship, it is — quite simply — much easier for the male partner to act in this capacity.

7

CLEANER EMPLOYMENT IN CONTEMPORARY BRITAIN

As we showed in Chapter Two, the cleaner constitutes the most important category of waged domestic labour in contemporary Britain. In this chapter we first outline and illustrate the basic characteristics and forms of cleaner employment, looking at day-to-day work practices and the terms and conditions of cleaner employment. As will become clear, these are very different from those found within nanny employment. We then examine the social relations found within cleaner employment. Whilst these relations are shown to have considerable similarities to nanny employment, some critical differences occur – particularly in those instances where both a cleaner and a nanny are employed. We conclude the chapter by accounting for these social relations.

INTRODUCING CLEANERS AND THEIR EMPLOYERS

In the course of our fieldwork we encountered two forms of employment of cleaners. The first was where one cleaner worked for one household. The second was multiple cleaning; where one cleaner worked for more than one household, usually two, but sometimes as many as three or four. In this respect, the cleaner form of waged domestic labour is more straightforward than the nanny. However, beyond this, private domestic cleaning is complicated by the total number of sessions per week which a cleaner performs for a household. No clear trends emerged with respect to the dominant form of cleaner employment, although it would be fair to say that working for one employer (either once a week or more than once a week) was more common than cleaning for more than one employer. But, given the small number of cleaners interviewed and the way in which cleaners move back and forth between single and multiple employer cleaning, we would not wish to be categorical about this. In this section we outline the two basic variants of cleaning work, as well as the day-to-day work activities of our respondents. We then move on to examine how the number of sessions performed per household per week mediates the nature of employment. As with our investigation of the nanny, we begin our discussions with portraits of the various forms of cleaner employment.

Mary (working for one household once a week)

Mary is 46 and married to a self-employed painter/decorator. Unusually amongst the cleaners we interviewed, Mary had taken O levels, done business studies at the local technical college and then worked for an insurance company, where she participated in an insurance training scheme. She remained in insurance work until she had her first child. Divorce was the main factor which pushed Mary into cleaning: as a single parent dependent on benefit, she found herself able to work only in circumstances which fitted in with her childcare responsibilities. Now remarried, Mary has moved in and out of various cleaning jobs in accordance with personal circumstances. Currently she cares for her invalid mother but works once a week (for four hours) for the Smiths, for which she receives £10 cash-in-hand. She has worked for them for eighteen months. Celia Smith is a manager of a large retail store and John Smith a merchant banker.

Doris (working for one household five mornings a week)

Doris is another woman whose route into cleaning reflected caring responsibilities associated with kin. When she started private domestic cleaning, Doris quickly expanded the number of households for whom she worked, to the point where she was supplementing benefit to the maximum level permitted by Department of Social Security regulations. Since then, however, other caring responsibilities have intervened to shape her labour force participation. As her mother has aged, Doris has found it increasingly necessary to perform domestic labour for her. Finding that she was struggling with the burden of five sets of domestic labour (her own, her mother's and that of three employing households), she gave up two of her employers. Currently she works five mornings a week (a total of twenty hours a week) for one household, for which she receives £48 cash-in-hand. Her employers are both doctors and also employ a nanny – Louise. Doris has been working in this manner for just over a year.

Sheila (working for two households on a once-a-week basis)

Sheila has been working as a cleaner for four years. Her move into cleaning resulted from her husband's ill health and the household's consequent reliance on invalidity benefit. Initially Sheila worked for two mornings a week for a firm specialising in household cleaning services. Through this job she met Linda Green. Not unusually amongst our cleaner respondents and cleaner-employing households, Sheila was 'poached' by Linda Green about six months after she had begun being her regular cleaner. She has worked in an independent arrangement for the Greens for over three years, receiving £12 for four hours' work. In addition, Sheila also works for the Tylers, again for four hours once a week. For this she receives £16. Alan Tyler runs his own business and Belinda Tyler is a teacher.

Brenda (working for four households on a once-a-week basis)

Brenda is 46, a divorcee and now living with another partner. She has a teenage daughter. Since leaving school she has had a variety of jobs, many of them in factory assembly. She started private domestic cleaning a year ago on the suggestion of a friend. Her partner is in receipt of income support. Currently Brenda works four mornings a week for four different households, none of whom know one another. Of the four we only interviewed one. Ruth Vine is a chartered surveyor and her husband (Charles) is a university lecturer. Brenda is paid cash by all her employers and accumulates £40 for her week's labours.

The above portraits highlight some important differences between the cleaner form of waged domestic labour and the nanny. Beyond the obvious part time/full-time distinction, the most significant of these differences is that employers have little control over when – or indeed, if – a cleaner works for them. Instead, it is the cleaner who shapes the degree of labour force participation; and this can be as part-time as cleaners want to make it. At one extreme, for example, is Mary who works four hours per week, while at the other are Brenda and Doris who both work approximately twenty hours a week.[1] Moreover, it is clear that cleaners expand and contract the number of their employers to suit personal circumstances. Doris is a classic example of this. Furthermore, whilst households share cleaners, the form of sharing is one shaped by cleaners and not by employing households (compare the nanny-share). A second difference between the cleaner and nanny forms of waged domestic labour is the location of private domestic cleaning within the informal sector. Cash-in-hand, involving small cash payments, and non-contractual, this is a long way from the employment terms and conditions for work as a nanny. The third difference between the nanny and the cleaner is the nature of the work and its mediation by the number of sessions performed per employing household per week.

Cleaning tasks themselves are ones which Oakley (1974) has likened to routine, repetitive industrial work and, at their simplest, involve

Table 7.1 Basic and secondary cleaning tasks

	Basic/essential tasks	*Secondary tasks*
	Dusting	
	Vacuuming	
One session per week (3–4 hours)	Kitchen cleaning i) surfaces ii) floors iii) oven(s)/hobs iv) cupboards	
	Bathroom cleaning i) bath/shower ii) basin iii) toilet(s)	
	Bedrooms	
	Windows (insides)	
Two sessions per week	Washing Ironing	Furniture polishing Floor polishing Silver/brass Ornament cleaning

Source: Fieldwork interviews

the following activities: dusting; vacuuming; kitchen and bathroom cleaning; bedroom cleaning and cleaning the insides of windows. As might be anticipated, we encountered little variation between individual cleaners in their performance of these basic tasks. Indeed, most had evolved a routine of the type described by Mary: 'I start with the hall. Then the kitchen, then the lounge downstairs. Then the four bedrooms and the two bathrooms upstairs. So I do all the downstairs first and then go upstairs.' This routine seemed to have been derived from their own way of doing things at home, though the basic pattern was complicated in those circumstances where a cleaner worked for one household on more than one occasion per week.

Table 7.1 details in full the distinctions which exist between single-session cleaning for households and multiple-session cleaning. In multiple-session cleaning (as opposed to cleaning for multiple employers) the tasks outlined above are supplemented by other basic tasks, as well as by secondary tasks. Interestingly, the other basic tasks *do not relate to cleaning*. Instead, they are the other labour intensive, highly gender-segregated tasks of household washing and ironing. Secondary tasks, although cleaning-related, are more specialist and/or time consuming activities, as well as irregularly performed. These include polishing furniture, cleaning ornaments, polishing silver and brass and polishing floors. In its multiple-session form, then, private domestic cleaning extends well beyond the basic household cleaning tasks to include activities which are outside the cleaning remit altogether. We consider in depth later in this chapter how and why this expansion in the activities of the cleaner occurs. However, for the moment it is important to stress that in practice, and much as with the nanny, the division of labour upon which the cleaning form of waged domestic labour is based can break down. Where multiple-session cleaning occurs, the cleaner becomes the cleaner-cum-ironer-cum-washerwoman. Much as with nannying, cleaning in contemporary Britain involves considerable elision with the activities of the 'housewife'. Before accounting for this elision we examine in depth the characteristics of cleaning work, highlighting as we go the major differences between it and nanny employment.

AUTONOMY AND DIRT: CONSTRUCTING SELF-WORTH FROM OTHER PEOPLE'S MESS

In exploring the work of the nanny, we established fairly early on the central significance of trust, consensus and negotiation, both to the day-to- day reproduction of nanny employment and to its construction on the lines of mother substitution. The employment of cleaners is rather different. Notwithstanding the importance of trust within such employment, it is a form of work over which cleaners have a far greater degree of control than do nannies. Negotiation and consensus, although present on occasion within cleaner employment, are generally absent from day-to-day practices. Indeed, for most of the time the cleaner simply gets on with 'her own thing'. She is, in short, in a far more autonomous position than the nanny.

Day-to-day practices

Two spheres of cleaner employment show the high degree of worker autonomy experienced. These are the nature and organisation of the work itself and the terms and conditions of employment.

Some idea of the degree of autonomy experienced by cleaners is evident from the previous section, where we indicated how cleaners construct their own work routines, and how they base these on what they do in their own homes. Such comments were confirmed not just by our own observations but by the comments of cleaner employers. The comments of Pauline Collins are typical in this respect. They also contrast with her direct intervention in the work of the household's nanny.

211

INTERVIEWER: So, if we move on to the cleaner. You've had several cleaners. Are they all similar in terms of the tasks that you've asked them to do?

PC: Yes – vacuum, dust and just keep the place clean. I ask them to concentrate on the rooms which have the most traffic and use. My very first cleaning lady – sadly she left because she became ill, otherwise she'd still be here. She was very genteel. *There were things that she wouldn't do though.* Then the next cleaning lady. She was also in her fifties, and she was a real whiz. She cleaned furiously! She was a real tornado! She'd defrost the fridge, clean the oven – great big jobs, *without being told. She organised me.* And now there's Mrs Jones, who's also in her fifties. She's not as good as Mrs Black, but *I leave her to get on with it.* And I know that the kitchen will be done and the bathroom and the loos. (Our emphasis.)

Beyond making initial general directions, as Pauline Collins' remarks show, *the employment of a cleaner is about being organised by the cleaner.* Work routines are structured by cleaners and cleaners even dictate what forms of work they will (and will not) do, as well as what materials they prefer to use in their work. Here, for example, is Sarah Davies on this theme:

> I was vetted by the cleaner and so was the house! And, after a nerve-racking ten minutes she decided that she could come. But she does bring all her own dusters and her own iron. She tolerates our [vacuum] cleaner! And she buys whatever cleaning materials she feels we need and she leaves a note for me to pay her.

Although the norm, the above pattern of organisation is occasionally over-ridden by employers. Indeed, most of our cleaner-employing households referred to leaving notes for their cleaners if they wished them to depart from their normal routine. Here, for example, is Janet Jones:

JJ: I might leave her a note. Say I might ask, 'Can you spruce up the spare room a bit because I've got a guest coming to stay?'

Such interventions, though, are very much the exception. For the most part, and unlike the nanny, the cleaner shapes the nature and organisation of working practices.

Employment terms and conditions

If anything, the terms and conditions of cleaner employment provide an even better indication of the autonomy of the cleaner than the

212

organisation of the work itself. As the previous examples of both Mary and Doris illustrated, private domestic cleaning is an occupation characterised by frequent (and easy) employer additions and subtractions. For many of our respondents, the early stages of private domestic cleaning had involved work for several households, but the physical burden of this workload undoubtedly proved to be too much for many — particularly for older cleaners, and for those such as Doris with other commitments. The response in many cases therefore had been to 'finish with' particular employers and keep on others, that is to reduce the number of employers, but to do so by their own choice, rather than that of the employer. Beyond this, our respondents suggested that they were also very much in control of when they worked for particular employers. A classic example of this is Iris, whose daughter at the time of our contact with her was expecting her second child. Iris clearly anticipated taking time off work to go down to assist her daughter in the immediate post-natal period:

> I'm going down to me daughter's in January, 'cos that's when the baby's due. And I'll just leave a note, and I'm back whenever I'm back.

Others, such as Rose and Sheila, also took temporary breaks from work as and when it suited them. Thus Rose was shortly to go to stay with her daughter for three weeks, whilst Sheila took numerous holidays throughout the year:

> When I want to be off, I just say, . . . I'm me own boss like. If I'm on me holidays I just leave a note – 'On holiday, I'll not be in.' Or I just phone up and say, 'I'm away on me holidays.'

Moreover, cleaners even exhibited a considerable degree of control over when they worked for a particular employer within any one week. Obviously, such flexibility was far more difficult (if not impossible) for those engaged in the extreme forms of cleaning for multiple employers, but those working for one or two employers had a considerable degree of flexibility, and were able to shift work commitments to fit in with events such as medical appointments. They were also able to use personal commitments to alter their terms of employment. Thus, Sharon, who has a young daughter, has recently altered her work times:

> It's usually three hours, but *it'll have to be two hours on a Friday now* [our emphasis] because I don't get here till nine-thirty and I'll have to leave at half eleven to pick her up [from nursery].

Such representations were confirmed by employers. Here, for example, is Janet Jones again:

INTERVIEWER: So are there any times when she doesn't come?

JJ: Sometimes. For instance, she might say, 'I've got to go and clean for so and so because they're in a lot of hassle.' I've had a period of six weeks when she didn't come. For two of those weeks I was on holiday. The rest of the time she'd got other things she wanted to do. She works most weeks but it's flexible.

Important as all this is in demonstrating cleaner autonomy, there is one further sphere of control over employment terms and conditions which cleaners exercise which is of even greater significance. This relates to the conditions within which they are willing to work. There are very definite boundaries with respect to private domestic cleaning, beyond which cleaners refuse to work. These relate to the levels of human dirt which cleaners are prepared to tolerate, although evidence suggests that 'acceptable' conditions are also mediated by the gender of potential employers. There are some circumstances in which private domestic cleaners *will not* clean. Indeed, all twenty of the women we interviewed either refused point blank to work in conditions of intolerable dirt or left very quickly. As Sheila's and Brenda's comments make clear, intolerable dirt is equated with muck and filth:

Sheila: You have never seen anything like it in your life! The carpets were sticking to the floor. It was disgraceful. The banister rails were all finger marks off the kids. And the walls! It was filth! Fred came to pick me up and I said, 'Fred, it's a pig sty. I'm not going back.'

Brenda: I had another flat in X. And they were students! Dreadful – filthy! You wouldn't have believed how filthy they were! I was there only three hours and they expected me to do the whole place. And the ironing! Huh!! I left after two weeks.

To expect a cleaner to 'muck out', therefore, was beyond the bounds of the permissible. Instead, all of our respondents expected to work in conditions of tolerable (and sometimes apparently non-existent) dirt. However, they revealed quite definite expectations and dirt tolerance thresholds towards certain types of household. Doris, for example, had been prepared to clean for the following set up:

I knocked on the door and I said I was looking for Mr Brown. So he says, 'Yes come in.' And I've never seen anything like it in my life! They were students. There were more holes in the carpet than carpet. The three-piece was dropping to bits. If you did under the bed you'd be fetching out about half a dozen gin bottles! He was very nice with it. He was very young. I did my best to keep it tidy and to keep the windows clean.

Such comments provide a stark contrast to those of Sheila and Brenda above, and it is hard to avoid the conclusion that Doris was only prepared to tolerate such working conditions because these employers were male and (manifestly) incompetent and incapable. In short, they conformed to Doris's expectations regarding the domestic capabilities of men.

All this, fairly obviously, is at some considerable remove from the situation discussed in Chapter Six. But why? Why should private domestic cleaners, themselves 'unskilled', with minimal (and usually no) qualifications, and largely in-and-out work histories, exert such a strong degree of control over this form of waged domestic labour, whilst nannies (the trained, qualified professionals) do not? In the previous chapter we showed how the ideological construction of motherhood and mothering shapes nanny employment. Here we argue that the social connotations which pervade the activity of cleaning ensure that this form of waged domestic labour is characterised by worker autonomy.

At this point in the development of our argument it is necessary to consider the connotations which surround the activity of cleaning. Thus far in our discussions we have stressed that cleaning is hard, physical labour. We have emphasised that the work is at the bottom of the female occupational hierarchy. Furthermore, we would agree with Oakley's drawing of analogies between cleaning and routine, repetitive industrial work – particularly where women are cleaning for a number of house-holds. However, what has been missing so far from this representation is an acknowledgment of the meanings which surround cleaning and which are used to construct it as an occupation. We now show these to be critical to the construction of worker autonomy within cleaning.

Without doubt, cleaning is an occupation to which considerable social stigma is attached, and this was recognised clearly by our respondents, both implicitly and explicitly. Brenda, for example, in characteristically forthright terms, categorised her work as 'the lowest of the low'. Others were less damning, but there was little doubt that they would have accepted Brenda's representation. The following comments from Doris, for instance, although referring for the most part to her employment history, contain a definite subtext regarding the stigma associated with cleaning:

D: . . . before that, when he came out of the army, I went back into nursing I got onto the SEN [State Enrolled Nurse] course and did a month's training on the wards. Me mother was minding the kids at the time you see. John had started school, but Kevin hadn't. And I think she was getting a bit sick, because me mother's attitude was 'You should be bringing them up'. And, of course,

Bob was getting cheesed off because me mother was always complaining. And when I was coming in I was having to study. So I just packed it in. But if I had my time again, that's what I'd want to do. . . . I've never really got to do what I wanted to do. Even now.

INTERVIEWER: Do you regret that?

D: Yes I do. I mean I wouldn't change being married to me husband. Otherwise I'd never have had the pleasure of the family. I wouldn't have had me sons, me daughter and the grandchildren. But it's what you could've done isn't it . . . I can't say that I enjoy it. I do it because I need the money. I need the money so I do it. And I'm too old now to go out and train for something else. So I've got to make the best of it.

The strong social stigma attached to cleaning is, undoubtedly, a reflection of its direct association, not just with dirt, but with personal, bodily dirt and the taboos which surround this. Ungerson (1983), for example, has shown how tasks associated with the removal of human dirt have become gendered as female work, and has stressed the existence of strong social taboos over men doing such work. But, insightful as such comments are, they do not begin to tackle the question of how those who clean cope with the stigma. We are in no doubt that women who work as private household cleaners do so through developing powerful concepts of self-worth and self-respect. We believe worker autonomy to be central to such self-concepts. Indeed, in controlling who they clean for, when, what things they will and will not clean, and even in what conditions they will clean, private domestic cleaners are able to counter at the level of the self the social construction of the cleaner as 'the lowest of the low'. So, if we are to account for the greater degree of worker autonomy of cleaners relative to the nannies, we need to recognise that this says a great deal about the social meaning of personal, bodily dirt in contemporary British society. The cleaner's autonomy is not just about the relative values of childcare versus material things. Rather, the only way through which the private domestic cleaner can retain self-respect within an occupation which is continually about clearing up other people's mess is through creating autonomy in the workplace. Moreover, in making minimal demands on their cleaners, employers appeared to recognise, albeit largely tacitly, the importance of this autonomy and self-respect. In addition, in clearing up before their cleaners start work (an activity which a number of the employers referred to), they also seemed to be recognising the centrality of dirt and dirt thresholds to the reproduction of cleaner employment. In their different ways, then, employers and household cleaners negotiate and reproduce the meanings which pervade personal dirt to produce a form of waged domestic labour in which – in contrast to the nanny – it is the cleaner who *appears* to hold all the cards.

Having examined the day-to-day social practices and terms and conditions of cleaner employment, we move on to look at the social relations of this particular form of waged domestic labour.

PART OF THE FAMILY / DISTANT EMPLOYEE: THE TWIN FACES OF CLEANER EMPLOYMENT

In discussing the social relations of nanny employment, it was clear that these are relations constructed between women. Men's involvement in nanny employment was shown to be marginal and generally confined to decisions regarding employment and disciplinary issues. If anything, the social relations of cleaner employment are even more strongly constructed as relations between women.

This was visible at various stages of our fieldwork, most notably in the social practices of cleaning (whether observed or described) and in the representations of employment by both cleaners and their employers. It is, for example, the *female* partners within our interviewed households who were 'vetted' by cleaners; it is the *female* partner for whom the cleaner leaves notes, and it is the *female* partner who – on occasion – leaves notes for the cleaner. In addition, as Mary explains, it is the *female* partner who (usually) shows prospective cleaners around: 'She showed me round the house and told me what she wanted me to do.' Moreover, it is usually the *female* partner who arranges the payment of the cleaner (i.e., where the money will be left). Not surprisingly, these practices were replicated in our more abstract discussions concerning employer–employee relations. Here, for example, is Christine Uttley on Avril:

INTERVIEWER: Who do you think Avril sees as her employer?
CU: Oh that's definitely me.

And Avril on Christine Uttley:

> I leave her a note and say I've run out of polish and bathroom cleaning stuff. I just leave her a note saying get it for me the following week.

Barbara Roberts and Sharon articulated much the same ideas:

INTERVIEWER: On a day-to-day basis, whose responsibility is Sharon?
BR: Mine. Do you mean in terms of payment?
INTERVIEWER: Yes, and generally too.
BR: Mine.

To Sharon:

INTERVIEWER: All your arrangements are made through Barbara then?
You don't deal with Clive?

S: No not really. Sometimes he'll say to me, 'Can you tidy
up the study'. But not very often.

This pattern was repeated through our interviews. Moreover, the responses of both employers and cleaners were unhesitating and clear cut. There appeared to be no doubt that it was the female partner in a dual-career middle-class household who fulfilled the employer role. Unlike the situation within nanny employment, where households (although not nannies) tended at least to pay lip service to the involvement of male partners, the employment of cleaners was an uncontested, heavily gendered terrain.

We consider two reasons to be behind this situation. The first is cleaners' own expectations regarding domestic work. The second is the heavily gender segregated nature of cleaning tasks within dual-career middle-class households themselves.

In our discussions with cleaners, it became clear that these women had been encouraged to see themselves primarily as wives and mothers (that is, as unwaged domestic workers) and that they had spent much of their adult lives both performing and, for the most part, accepting this gender division of labour as 'the natural order of things'. In their eyes, therefore, domestic labour is a world to which men are – at best – only tangentially related. Indeed, whilst it was revealed that their own men assisted in a few ways with domestic work, we were left in no doubt that it was either the traditional rigid or the traditional flexible form of the domestic division of labour which prevailed in their own homes. The importance of this with respect to cleaner employment lies in the way in which this 'natural order of things' is assumed to transfer to the world beyond. In other words, cleaners themselves are purveyors of normative expectations regarding domestic work and its organisation, so much so that it was assumed that such patterns of organisation prevailed within the homes of their employers. Given this, *to work as a cleaner within any household comprising both a male and female partner comes to be conceptualised by cleaners as working for the female partner*. Brenda provides a classic illustration of this conceptualisation:

INTERVIEWER: And do they explain why they have one?

B: Well yes. One of them – the lady works. She works during
the day and I'm sure she works at night. In fact, she never
stops. Now Mrs Smith she's a headmistress. Her house is
the most immaculate house. And she does a lot of
entertaining. She has a lot of people staying. And again

> she works full-time. And her husband, he works full-time. They are really, really nice people. *She* just hasn't got time [our emphasis].

Brenda therefore sees herself as working for Mrs Smith (and three other women), because Mrs Smith hasn't got the time to do the household cleaning. For her, Mr Smith just doesn't come into the equation.

At the same time, the dual-career middle-class household itself is as formative in constructing the social relations of cleaner employment. Thus, whilst these households tend to share certain domestic tasks, other tasks – particularly the labour intensive, regular and physically demanding ones (i.e., cleaning, washing and ironing) – are overwhelmingly constructed as 'women's work'. As such, it is all too easy for the employment of a cleaner to be shaped as if it is the female partner, and not the household, for whom the cleaner works. Many of the female partners within our employing households articulated these sorts of associations when asked to suggest explanations for why the employment of a cleaner is constructed as between women. As Barbara Roberts explained:

> I suppose [it's] really because I'm there more often and she's involved in the domestic side of things, which Clive isn't at all. And well, I suppose it's because she's a woman she's my responsibility.

At one level, then, the social relations of cleaner employment are very similar to those within nanny employment. Like nanny employment, the relations are ones between women. Moreover, both the gender division of labour and associated normative prescriptions regarding household cleaning and childcare operate to construct this form of waged domestic labour on heavily gendered lines. However there are some critical differences between the social relations of cleaner and nanny employment which it is important to emphasise.

In Chapter Six we demonstrated the central importance of the social relations of false kinship to nanny employment. In contrast, though the employment of cleaners *can* exhibit relations of false kinship, such relations are far from inevitable. Indeed, their development is very much facilitated by the *absence* of the nanny form of waged domestic labour. The Uttleys, Tylers, Keiths and Isards illustrate the types of differences to which we are referring here. The first two households employ *only* a cleaner. The second pair also employ a nanny.

The Uttleys (Avril's employers)

CU: I don't really consider Avril to be a cleaner. . . . She's really one of the family rather than an employee.

JU: It's a strange sort of relationship really. I mean she comes round here socially as well with her husband, and we swop plants and things. Things that a cleaner wouldn't normally do.

CU: She can get away with asking the boys to do things that I would never do. She is more like a granny to them . . . she used to bring them things when they were ill and things like that. It really isn't an employer–employee relationship. . . . If I was being really truthful I would probably say that she is like a granny to the boys and we look upon her like that as well. In that respect she's part of the family.

The Tylers (Mary's employers)

CT: I must say I don't think we have a conventional employer–employee relationship at all really. It works very very well. It's a working relationship which hasn't got any clearly defined parameters that you tend to work within. It's a very flexible one with give and take on both sides. Like I said to you, she probably is my best friend. She would jump to help me any time for any reason. I know that. And not just for cleaning. I know that she would come to see that everything is alright when we're away for instance. That's beyond the employer–employee relationship. It's a friendship thing.

The Keiths

INTERVIEWER: So can you describe your relationship with her?

MK: It's difficult. Very difficult.

INTERVIEWER: She's not part of the household then, like Lizzie?

MK: No, not really. She resents Lizzie because she was there before Lizzie came. . . . Also I think she resents Lizzie because we have become such good friends. She's invited to christening and birthday parties and she isn't. So she isn't really part of the family.

The Isards

JI: She [the cleaner] used to be a childminder before, so I can't understand why she's cleaning now. I don't really know why she's doing it. And I don't have the same sort of relationship with her as I do with Jo [the nanny] because I don't really see her. So really I know very little about her. All I know about her is that she has two grown-up daughters and that came out when I interviewed her. And she lives in A. And I don't even have her phone number.

220

I think I'd probably be, or try to get, closer to her if Jo wasn't here.

Cleaners too articulated these differences. Here, for example, Avril confirms Christine Uttley's representation of her 'grandmother' status:

If the little one is there he thinks he is helping out you know. The other day I said something to him. He didn't take any notice. He just treats me like another gran.

Mary provides another instance of familial incorporation:

If I go up there [to the Tylers] and Belinda's off work – she's a teacher. She'll say, 'Leave your coat on.' And I'll say to her, 'Where are we going?' And she'll say X. I'll say, 'But I've no make-up on.' She'll say, 'Put some of mine on.' You know. We go to X for the day. And she still gives me my wage. ... And she has two little kiddies. And now and again I buy them something. Just novelties. Like slides for their hair. And when we go shopping I'm their Auntie Sheila.

This contrasts markedly with Brenda's comments, from which it is clear that she has difficulty with the concept of her employers as 'friends' (let alone as family):

Well, they're not really like a friend, except for one. But then, I guess I still wouldn't say that she's a friend. If I was in trouble I wouldn't go to her like. And the others. It's not like when you're employed in a factory. They're not that bad. In fact they're nice. They're more friends than employers. But I wouldn't say they're friends.

In what follows we examine the nature of these two forms of social relations in more depth, before moving on to look at their implications.

There are considerable similarities between those instances of the employment of cleaners which exhibit false kinship social relations and the circumstances we discussed in relation to nannies. The cleaner sees herself as 'helping out' the female partner, rather than as a waged employee. Consequently, waged work takes on the status of a favour and, very quickly, extends beyond the remit of cleaning. As a consequence, almost all those instances which exhibited either false kinship or friendship relations were ones where multiple-session cleaning occurred, and where cleaners who had started off doing the cleaning, now also did washing and ironing, house-sitting, plant-watering, pet-feeding and so. The comments of Christine Uttley and Celia Smith, both of whom have 'cleaners' who also do the washing and ironing, are typical in this respect:

INTERVIEWER:	Do you ever ask her to do anything extra?
CU:	Not really. She just tends to take things on and do them if she sees it. . . . She washes things here and then takes them home to iron them.
INTERVIEWER:	Do you pay her extra for that?
CU:	No she wouldn't take that.
INTERVIEWER:	So you've offered then?
CU:	Oh yes. She say's she is doing us a favour.
INTERVIEWER:	Does she ever do anything that you would consider to be extra?
CS:	Yes. Off her own bat she will sometimes. If she sees I've got an enormous pile of ironing, which happens from time to time, she'll just take it home – just off her own bat – and she'll try not to let me pay for it. She says she does it as a favour.

Beyond this, such households frequently received 'gifts' such as flowers and vegetables from their employee's gardens.

What is going on here is the construction of cleaning as a form of work which lies beyond exchange relations. Instead of the wage relation, it is 'helping out', 'favours' and 'gifts' – themselves standard features of kin and/or friendship relations – which constitute the underlying structuring principles of this type of employment. However, it is vital to appreciate too that – at least in the short term – reciprocity (the norm within kin relations) is largely *absent* from this formulation. Employers are *not* expected to reciprocate favours by cleaners, and – most certainly – are *not* expected to pay their cleaners for these 'extra' services. Indeed, to do so, as Christine Uttley for one recognised, would be to misunderstand the entire basis of the relationship. Not only might such attempts be conceptualised as an insult, but – if payment were offered – it would signify an attempt by employers to reaffirm the relations of exchange. The logical extreme of this complex situation is articulated by Sharon: 'I'd probably clean for Barbara even if she didn't pay me because she's so good. We're really good friends now.'

In complete contrast to these arrangements are those where the wage relation assumes greater importance. In such circumstances, whilst cleaners still see themselves as 'helping out', any notion of favours is absent, and gifts were definitely ruled out. Not surprisingly, Brenda (with four employing households) provided the clearest example of this alternative construct. Indeed, in the following extract we can see how she resents the expectations of one of her employers regarding 'extra' cleaning services:

INTERVIEWER:	So can you just describe what you would do.
B:	Always the hoovering. I'd hoover from top to bottom.

> And polishing Some of them want things done with
> proper polish. The one in X – she's a bugger that one!
> She has us polishing this awful floor. I could murder her
> for that!

For Brenda, then, and indeed for all the women who had engaged at
some time or another in working for multiple employers, cleaning was
conceptualised *as a job*. It was a way of making money. Construed thus,
and uncomplicated by the web of false kinship and/or friendship, there
was no way that Brenda would clean without payment. Indeed, in her
view it was the wage relation itself which inhibited the development of
friendship-based relations within cleaning. Employers – she felt – could
not afford to become too friendly with their employees:

INTERVIEWER: Why's that?
B: Because you may have to give them the sack. Or you may
have to finish with them for whatever reason. And
whatever, you're paying their wages. You're the piper.
You can't really afford to get that friendly.

In certain circumstances, then, the employment of cleaners is characterised 'by love, not money'. In other instances, it most definitely is
shaped by the wage labour relation. In the latter context, the terms
and conditions of employment mean that cleaners and employers
encountered few of the problems discussed in relation to the nanny in
Chapter Six. However, for those whose relations are bound by the ties
of false kinship/friendship, it is a familiar story.

For cleaners, pay is the only major problem area within a form of
waged work structured on false kinship lines. Paid at an hourly (i.e.,
casual) rate, many of the cleaners whom we interviewed had not had a
pay rise since they started working for their employers. Moreover,
asking for a pay rise in the context of a form of work which they had
structured (and conceptualised) as predominantly outside exchange
relations, was extremely difficult, if not impossible. As Sheila commented:

INTERVIEWER: You were saying earlier that pay is an issue. How do you . . .
S: How do you put in for a rise. Exactly!!
INTERVIEWER: Are you going to broach that?
S: Like Fred says, 'You should really ask.'
INTERVIEWER: Is that difficult?
S: It's very difficult to ask. You cannot go and say, 'Look I'm
putting in for a rise. I need a rise.' But as Fred says, 'You
end up working for less.'

In a sense then, cleaners who construct their employment in terms of
false kinship and/or friendship relations are caught in an unintentional

pay trap of their own making. They become entirely dependent on their employers for pay rises, yet many convey the impression to their employers (through the existence of favours and gifts) that pay is a minor concern to them.

For employers, too, the false kinship/friendship web is a difficult one to negotiate and, indeed, many employers were well aware that they were bound up in a relation of long-term 'debt' repayment. Employers, therefore, were aware that cleaners were individuals to whom they had certain responsibilities and obligations – that the construct of 'helping out', although employee-generated, eventually 'cut both ways'. Thus, we encountered several instances of employers continuing to employ – or at least continuing to pay – cleaners well beyond the time when their usefulness as a cleaner had passed (either because of changed circumstances in the employing household or because the cleaner was too old/infirm to carry out the job properly). In one case, the household even employed another cleaner on another day to clean their house, whilst still retaining their old cleaner and continuing to pay her. The Uttleys provide a good instance of the type of obligatory responsibility which employers could feel towards cleaners:

INTERVIEWER: Would you still pay Avril?
CU: Oh yes. I think she relies on the money from cleaning jobs to keep going.
JU: Avril's done a great job for us over the years. There were times in the past when we really couldn't have got by without her. So, if it comes to the point where she needs us more that we need her (which I suppose isn't too far away) we're not going to abandon her. I think after so long we both feel that we have a certain amount of responsibility to her. I would feel really guilty if we abandoned her.

The social relations of cleaner employment then *can* be characterised by relations of false kinship. But they can also be structured on the lines of the more distant employer. We now examine how and why these differences occur, as well as their relation to nanny employment.

Figure 7.1 portrays the two types of social relations which we encountered within the cleaner form of waged domestic labour. The first has clear parallels with the relations depicted in Figure 6.1, although it is important to stress that the pattern of familial incorporation here is more one-way than two-way. Within cleaner employment structured on false kinship/friendship lines, it is almost invariably *the cleaner only* (and not the cleaner's family) who assumes the status of psuedo-kin/close friend. Moreover, we found very little evidence to suggest that either the adult members of employing households or

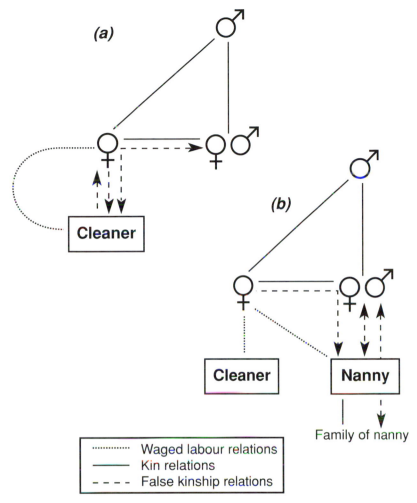

Figure 7.1 The social relations of cleaner employment
Source: fieldwork interviews

(more particularly) the female partners within them, assume pseudo-kin/close friend status within the family of the cleaner. Instead, the latter were classified solely in terms of their relationship with the cleaner.

As with the nanny form of waged domestic labour, our explanation for the development of such arrangements stresses the importance of emotional 'debts'. Whilst the cleaner cannot be seen to enable the reproduction of the dual-career household form in the way that the nanny does, her employment is still enabling, and it is the recognition of this sort of enablement which lies behind the comments of Christine and James Uttley above regarding Avril. Thus, rather than allowing the female partner within a dual-career middle-class household to work full-time, the employment of a cleaner facilitates the availability of 'quality time' for the household. As we suggested in Chapter Four, this quality time could be used in various ways by partners. For example, in many households the employment of a cleaner was seen as allowing parents to spend more time with young children than would otherwise be possible, thus enabling the negotiation of working parenthood, and working motherhood in particular. Cleaners, in short, allowed many of the female partners to be mothers in the time–space beyond employment, rather than part-time housewives. Given this, it is hardly surprising that a number of the households felt some form of obligation and responsibility towards the cleaner.

Part of the explanation for the presence of false kinship/friendship relations within this form of waged domestic labour then lies with what such employment enables and facilitates. At the same time, though, whilst this enablement accounts for the feelings of certain of our employers towards their cleaners, it does not explain why their cleaners, in turn, construe these relations in similar terms. In order to account for this we have to return to the arguments in the previous section concerning the social meanings which surround personal dirt. As we stressed earlier, the social stigma associated with clearing up personal dirt is something which is countered by a significant degree of worker autonomy. Here we argue that, whilst conceptualising employers as 'friends' is yet another layer through which cleaners are able to build up powerful concepts of self-worth and respect, such conceptualisations are also a necessary part of this employment. As Ungerson (1983) has argued, the ties which bind women and personal dirt are ones which are constructed through the concept of caring. Moreover, Graham (1983) has suggested that these ties relate to specific people, rather than to caring in the abstract. In other words, if women are to work around personal dirt, then it helps if they care about those making the personal dirt! Cleaning in a private domestic capacity is clearly an occupation in which personal dirt is well to the fore. But, in being home-based, inextricably linked to personal, private space and tied necessarily to personal material objects, things and possessions, it is an occupation which is pervaded by the mess of the specific 'other' (and others). As such, it is not difficult to see how women who clean in a private domestic capacity are able to construct their work in terms of notions of personalised caring. Moreover, in constructing private domestic cleaning in terms of 'helping out', 'favours' and 'gifts', cleaners clearly move this form of work away from the terrain of waged labour and on to the terrain of unwaged domestic labour. In so doing, they allow this cleaning form of waged domestic labour to assume the characteristics of, and become equated with, the work which they do 'for love' in their own homes. It is, therefore, easy to see how (when shaped themselves by gender ideologies which encourage women to see cleaning up the personal dirt of kin as an expression of love) women who clean in a private domestic capacity can come to care about these specific others as friends and/or pseudo-kin. It is, therefore, the way in which women, caring and personal bodily dirt are bound together by gender ideology which accounts for the social relations of this form of waged domestic labour.

The alternative form of social relations within cleaner employment is one in which distance plays a key part (see Figure 7.1(b)), and it is important to emphasise that we only encountered such relations in circumstances where a nanny was also employed. Such relations were the product of the (usually unintentional) actions of the employing household (and particularly the female partner within this). For many of the households, a social relation characterised by distance was the unintentional product of the fact that cleaner and employer where rarely in the house at the same time. Unlike the nanny, then, the cleaner was someone with whom employers could have very little physical contact and/or interaction. Under these circumstances there may be little chance to develop relations grounded in false kinship/close friendship. Such is certainly the position articulated by Julia Isard at the start of this section.

However, there is little doubt that a number of cleaner-employing households exacerbated this situation through their actions (again probably unintentionally). Here, for example, is Julia Isard – immediately after the comments made previously: 'I pass messages to her through Jo.' Similarly, Pauline Collins again: 'I say to Julia, not that I mean to pry but, "Does Mrs Black seem busy all the time or does she have time to kill? Is she twiddling her thumbs?" In a way I want her to suss her out.'

In a number of households, then, the nanny is used as a direct substitute for employer presence, to monitor unseen waged labour. We can note, in passing, how such practices might work to reinforce the notions of mother substitution discussed in Chapter Six. However, there is little doubt that, in asking nannies to pass on messages which instruct cleaners and in asking nannies to check up on cleaners, employers create greater social distance between themselves and cleaners and, albeit inadvertently, a hierarchy of waged domestic workers within their own homes – one which reinforces the place of cleaning (and cleaners) at the bottom of the value scale within domestic work.

Although we consider relations of distance within cleaner employment to result from the unintentional actions of employers (and particularly of female partners) other features of cleaner employment may assist such constructions. One of these is the autonomy of this labour force. In circumstances where the cleaner appears to 'hold all the cards' it may be difficult for employers to 'get closer' to their cleaner. A second feature is the differential social values placed on children as opposed to material things. If children are present within any household, therefore, it is almost inevitable that the waged domestic employed in relation to their care will be seen as doing more valuable work than someone caring for material things. Finally, however, there are the social characteristics of the cleaning labour force. As we stressed in Chapter Five, cleaners are older, more solidly working-class than nannies. As such, and as a number of employing households pointed out, they are socially more removed from employers than nannies. Moreover, given their age, their potential for incorporation as pseudo-kin within any dual-career middle-class household is much less than with nannies. It is, therefore, not surprising in our view that, where pseudo-kin relations had developed between cleaners and nanny-employing households, these took the form of 'gran'/'auntie' *to the children*.

In those circumstances where distance characterised the social relations of cleaner employment, cleaners were well aware of their significance. Moreover, as the following comments from Doris reveal, they could not but be aware of their perceived value relative to a nanny:

> That did poke us. When I first started here, they'd said to me that when they got a rise, I got a rise. So I'd been here three weeks. And they were sitting at the dinner table and he said, 'Oh we've got some good news for you Louise, you're getting a pay rise in April.' And then he said to me, 'Oh I don't know about you.' So anyway, I thought, 'Oh he's having a bit of a joke.' So the next morning, when I came in, he said, 'Oh I've had a word with Marianne about your pay rise and we've decided that it would be best for you to be here for six months.' And I was poked about that because anywhere else you get a pay rise regardless of when you started.

The reactions of cleaners and nannies to such discrepancies seemed to be to create strong ties of allegiance. Thus, Doris regularly minds Louise's charges for her, whilst – in turn – Louise will help Doris with tidying and washing. Furthermore, in one instance we encountered the nanny and cleaner had worked out a complex form of non-verbal communication between them which spelt out the 'mood of the day' in the employing household! In practice, therefore, waged domestics, when employed together, tend to resist any prioritising of one form of waged domestic work (and worker) over another by employers, and see their tasks as interwoven and – to a degree – as mutually exchangeable. In seeing things in these terms they are not just resisting employer-constructed hierarchies but are resisting too the subdivision of domestic tasks on which the separate employment of the cleaner and the nanny are based.

SUMMARY

As we have shown in this chapter, being a cleaner in contemporary Britain is rather different to being a nanny. Far more autonomous, cleaner employment is less 'messy', less contradiction bound, than the nanny. It contains none of the elisions, for example, of childcare professional/mother substitute. The cleaner *can* be a much loved individual, someone incorporated within the web of familial relations – a giver of favours and gifts. But, alternatively, she can also be a much more distant employee; someone who simply and invisibly just gets the job done. Instead of being characterised by double constructions, cleaner employment appears to be characterised by alternative constructs. Why?

In accounting for the above we draw attention to two issues. The first relates to the contrasting employment status of the nanny and the cleaner. Here we highlight two points. First, cleaner employment is usually a casual form of waged domestic labour. Given this, and particularly in its single employer form, it exhibits strong contrasts with the full-time nature of nanny employment. As such, the potential for the development of powerful notions of waged labour is considerably less than is the case with the nanny. In such circumstances, therefore, the contradictory tension between the social relations of wage labour and false kinship may not arise. Second, in those circumstances where working as a cleaner assumes the status of a part-time job (i.e., multiple-employer cleaning), the fact that cleaners rarely coincide with employers can inhibit the development of false kinship relations. The very forms of employment within the cleaning form of waged domestic labour therefore are ones which themselves tend to produce either/or constructs, rather than contradictory tendencies and double constructions.

A second explanation for why the cleaner form of waged domestic labour exhibits alternative constructs, pertains to the ideas which suffuse it. As we emphasised throughout the previous chapter, the contradictory tensions and elisions which characterise nanny employment are themselves the product of the ideological constructs of mothering and motherhood, specifically of the concept of the working mother. In contrast, the cleaner is an autonomous worker, rather than a specific substitute. We have argued that this is indicative of the social meanings which surround personal, bodily dirt. In the one case, then, we have a set of meanings which enable one form of waged domestic work to be constructed in terms of polarities and distance. But, in the other case what we have is another set of meanings which compel another form of waged domestic labour to be constructed on the basis of close collaboration and consensus. The first produces a form of labour characterised by an autonomous worker (who may, or may not, use the construct of caring to build social 'ties which bind'), and the second a form of waged domestic work suffused with the tensions of mother substitution.

Ultimately, then, the different characteristics of the nanny and cleaner forms of waged domestic labour in contemporary Britain are themselves products of the nature of labour substitution within both forms of waged domestic labour and of the respective ideologies which shape these. However, whilst wishing to stress the differences between the two most important forms of waged domestic labour within contemporary Britain, it is also important to draw attention to their similarities. In particular we highlight the heavily gendered nature of both jobs, and the way in which the traditional form of both the gender division of labour and the domestic division of labour ensure that these forms of domestic labour substitution are constructed as between women. Moreover, we would stress that, whilst at one level the resurgence of both forms of waged domestic labour within the homes of the new middle classes points to the subdivision and separation of domestic tasks, at another these forms (particularly in day-to-day practice) exhibit strong parallels with the unwaged domestic labour of the housewife/mother. Given the centrality of domestic work to the gender identities of women in Britain throughout the post-war period, this in itself is unsurprising. However, it is important to appreciate that the co-existence of waged and unwaged domestic labour in contemporary Britain, and the tendency for the former to be constructed in terms of the latter, have implications which go well beyond the social relations and practices of waged domestic labour. We examine the broader implications of this resurgence of waged domestic labour in contemporary Britain in our final chapter.

8

THEORETICAL AND
POLITICAL REFLECTIONS

In this volume we have examined the resurgence of waged domestic labour in Britain through the 1980s and into the early 1990s; we have accounted for how and why this resurgence has occurred; and we have explored in depth the nature of the two main forms of waged domestic labour employed within middle-class households. We now step outside these concerns to examine the broader theoretical implications of this phenomenon in terms of class and gender, and consider the politics of waged domestic labour in contemporary Britain.

WAGED DOMESTIC LABOUR, CLASS AND GENDER

Without doubt, the resurgence of waged domestic labour in contemporary Britain constitutes a major change in the way class relates to social reproduction. As Glucksmann (1990) has indicated, the decline in domestic service in Britain from the inter-war period onwards brought with it fundamental changes in the relationship between the working-class and the middle-classes. The two classes were no longer in a direct employer–employee relation. As domestic service disappeared, so did its 'semi-feudal' social relations; and – most importantly of all – the reproduction of labour power became similar for both classes. Performed almost exclusively by unpaid women using mass-produced commodities, the only difference was the greater disposable income of the middle-classes giving them a purchasing power which, theoretically at least, could be allocated to the domestic sphere. Such arrangements persisted through much of the post-war period. However, as this volume has shown, the Britain of the 1980s saw the resurgence of waged domestic labour in the homes of the middle classes. Employer–employee relations have been re-established between the middle classes and working classes around domestic labour. New social relations grounded in both false kinship and wage labour have manifested themselves. Above all, fundamental class differences have been reasserted. No longer exclusively unwaged, domestic labour (or at least certain aspects

231

of this) within the homes of the middle-classes is increasingly being performed by one (and frequently more than one) waged woman, as well as by the unwaged labour of female partners (and sometimes male partners). And in at least its cleaning form, this waged domestic labour is being performed by working-class women.

Why this has happened is clear. On the one hand, the expansion in the professional division of labour, coupled with the growth in women's participation in service class occupations, has generated a crisis in social reproduction for middle-class households, particularly with respect to childcare. As we have seen, such households simply do not have the time–space to combine parenting with the demands of career-structured occupations, and, as such, they provide support for Gorz's thesis that the demands of 'work' for the middle classes are such that they are unable to perform reproductive labour (Gorz, 1989). On the other hand, it is also clear that the restructuring of the 1980s produced a crisis of a different order for many working-class households; one in which reproductive work within the homes of the middle classes was one of the few employment options available, particularly for working-class women. To a degree, therefore, we would concur with Gorz: the changing nature of 'work' through the 1980s in Britain has brought with it a fundamental reworking of the organisation of domestic labour on class lines. But, as our analysis has shown, there is more than the lack of time–space behind the employment of waged domestic labour within many middle-class households in contemporary Britain. Indeed, class-specific cultural redefinitions of leisure time, and the exclusion of various forms of domestic labour (notably cleaning) from this, proved as critical a precipitating force. This was no crisis in social reproduction in the first sense, but a challenge to the cross-class, post-war mode of organisation of domestic labour. Confronted by a conjuncture in which definite ceilings appear to have been reached with respect to the commodification of domestic labour (in that households still have to purchase the raw materials for cleaning), and by a massive expansion in 'leisure pursuits', many middle-class households are using their high disposable incomes to buy in replacement domestic labour for cleaning, legitimating this through the discourse of 'quality time'. This then is no crisis in getting the work done, but a redefinition of what forms of reproductive work partnerships are willing to perform.

As our analysis has shown, however, and again contra Gorz, the employment of waged domestic labour is not, as yet, widespread within the middle classes in general in Britain, but rather specific to a sizeable minority of such households (possibly as many as a third of all dual-career partnerships). We have explained this pattern in terms of intra-household arrangements. Particular forms of the domestic division of labour between partners have been shown to be highly likely to produce crises in daily social reproduction for individual households, especially following the onset of childcare responsibilities. Moreover, partner gender identities have been shown to be facilitative in enabling the employment of waged domestic labour. What we have not been able to do, however, is to examine the reverse side of the coin, namely why certain other middle-class households have *not* employed waged domestic labour in any form. This is clearly important, certainly in class terms: not only would it indicate the likelihood of the spread of employment of waged domestic labour within the middle classes (and hence the likely full extent of the reworking of reproductive work along class lines), but the existence of 'non-employing' middle-class households is, in itself, a powerful critique of Gorz's arguments.

The resurgence of waged domestic labour in contemporary Britain then offers some support for the Gorz thesis regarding the changing nature of 'work', the place of reproductive labour within this, and its mediation by class. It also points to the several failings in Gorz's argument, however. Unlike the previous form in which reproductive labour was differentiated by class in Britain (ie., domestic service), waged domestic labour does not – at least as yet – appear commonplace within the middle-class as a whole. Moreover, where it appears it does not relate to all forms of domestic work, and whilst some of these forms of waged domestic labour (childcare) reflect a genuine Gorz-style crisis in time–space availability for the middle-classes, others (cleaning) manifest changing evaluations of the various domestic tasks by partners. Beyond this there are a number of further implications of the resurgence of waged domestic labour which it is as well to spell out.

Perhaps the most important of these is the breakdown in the post-war cross-class identification of women in Britain with all forms of reproductive work which the resurgence in waged domestic labour signifies. For much of the post-war period in Britain, women's relation to domestic labour has influenced profoundly the form of their participation within paid employment. The traditional form of the gender division of labour and the culture of domesticity affected all women regardless of class. Now, however, the resurgence of waged domestic labour has broken – or at least recast – some of these associations. No longer physically responsible for all forms of domestic labour, many middle-class women are able to sell their labour power as 'honorary men'. Beyond this, there is clear evidence to suggest that in their waged form in Britain certain reproductive tasks are becoming closely identified with women from specific classes. Thus, whilst the daughters of white collar lower management and secretarial labour constitute a significant facet of the reproductive labour associated with childcare in the homes of the middle-classes, the messiest aspects of daily household reproduction are being transferred to working-class women.

Rather than all forms of domestic labour being identified with all women, a class-mediated hierarchy of domestic tasks is once more being constructed. At the top of the domestic task hierarchy are those tasks which are seen as pleasurable – childcare activities, cooking and shopping. These remain largely unwaged, tasks which middle-class male and female partners tend to share equally. Below this are the other domestic tasks – tasks such as those associated with the day-to-day care of children (bed-changing, washing, routine meal preparation, etc.) – which are increasingly being performed both by middle-class women and waged domestic labour. And at the bottom of the pile are the labour intensive activities (notably cleaning and ironing), which are increasingly being identified with waged domestic labour from the working class. As well as carrying the burden of reproductive work within their own households, then, working-class women in Britain are increasingly assuming part of the same responsibility within middle-class households, whilst it is the provincial lower middle-class who – in supporting their daughters 'in kind' (through non-market housing costs, meals, etc.) – effectively subsidise the childcare costs of their daughters' employers.[1] Meanwhile, domestic labour for the middle-classes who employ waged domestic labour has come to be about 'quality time' itself.

As well as signifying a transformation in the way in which class relates to domestic labour, the resurgence of waged domestic labour has major implications for gender, gender relations and their analysis. Here we have space to isolate only one such implication. This pertains to constructions of femininity.

In Chapter Three we highlighted the work of feminist historians on nineteenth-century domestic service in Britain, stressing the way in which they show domestic service to be central to the construction of class-based forms of femininity (Davidoff, 1983; Davidoff and Hall, 1987). In contrast, in the post-war period, with domestic labour being performed by working-class and middle-class women alike, it can be argued that domestic work played a consistent part in shaping constructions of femininity. Although varying no doubt in the ways in which they performed domestic tasks, working-class and middle-class women seem to have been equally exposed to post-war ideologies of domesticity and homemaking. Now, with the resurgence of waged domestic labour, such commonalities appear to be in the process of breakdown, if not quite collapse. As we have stressed, whilst middle-class women who employ waged domestic labour continue to perform certain domestic tasks, they do so in a way which alters dramatically their social relation to domestic labour. Thus, quite apart from the importance of false kinship relations in the day-to-day employment of waged domestic labour, we have seen how middle-class women assume the role of 'the manager' of domestic labour and domestic workers. Moreover, we have shown how the employment of waged domestic labour can lead to the segregation of domestic tasks. Such changes would seem to offer at least the possibility for the construction of new forms of femininity for middle-class women, ones in which the workplace (rather than the home) is foregrounded, and in which domestic labour figures both as an expression of middle-class women's organisational capabilities (ie., as reconformation of workplace skills/assets) and as a signifier of the home as a locus of pleasure.[2] In contrast, for those women who service the middle classes, performing domestic tasks for money (as well as 'for love') can only mean that domesticity and homemaking will assume greater prominence within their individual gender identities. Ultimately, the prospect which waged domestic labour raises is one which permits once more the construction of contrasting femininities (see Gregson and Lowe, forthcoming b and c).

As will be apparent from the tone of the above observations, we find it difficult to be positive in any way about the resurgence of waged domestic labour within contemporary Britain. In short, waged domestic labour signifies a highly traditional reconstitution of domestic work on class lines and opens the door to the construction of contrasting class-based femininities. Wherever one looks within contemporary waged domestic labour in Britain, one is confronted by a mire of difficult and divisive positions for women. These constitute the context for the political observations which follow.

And prospects for feminist politics?

Divisions between women constitute both the theoretical and political contradictions of feminism. If we are to avoid becoming trapped into contradictory strategies by our contradictory interests, then feminist politics has somehow to take these contradictions into account.

(Ramazanoglu, 1989: 174)

The resurgence of waged domestic labour in Britain through the 1980s and the reconstruction of domestic labour on class lines which this represents, is indicative of what Ramazanoglu refers to as the classical political impasse for contemporary feminism (and see too, A. Phillips, 1987). From being one of the key unifying threads in the lives of women in post-war Britain, reproductive work is increasingly being constructed on lines which divide women's interests, rather than unite them. In 'old' terminology, 'liberation' for some women is being achieved at the cost of the 'oppression' of others. A politics of fragmentation – of contradictory interests – seems to be inevitable. Or is it? In the final section of this chapter we consider this question in relation to the forms of waged domestic labour which we have considered in this volume.

Contradictory interests were never far from the surface throughout our interviews with both nannies and their employers, some of which became articulated in the pre-recession policy context of 1989/90. For nannies, it was the issue of pay which figured centrally in discussions. Unregulated, at the whim (or so it seemed to them) of the employing household, carrying with it heavy caring responsibilities and yet – for many – poorly remunerated in relation both to other (less responsible) occupations, as well as to employers' (supposed) earnings, pay was seen as the primary source of inequity. For employers, on the other hand, it was footing the bill with respect to childcare which was articulated as the strongest and most commonly cited grievance. In circumstances which were represented to us as ones of 'no choice' in relation to childcare provision (see Chapter Four), household members resented not so much the amount of household expenditure on childcare (indeed, many households stated that they wished they could afford to pay their nannies more), but the 'double taxation' which the 'above board' employment of a nanny necessarily entails. Seen thus, the interests of both parties are seemingly contradictory. For the one group, the obvious political strategy would appear to be organisation and mobilisation around wage regulation. However, the prospects here are far from encouraging. As various historical and contemporary studies have shown, the organisation of waged domestic labour and labour force characteristics means that political mobilisation through unionisation is extremely difficult, if not impossible, to achieve (see Chapter Three). In contrast, for employers, demands for financial assistance in meeting the costs of childcare constitute the clear politics of self-interest. Prior to the recession of the early 1990s, it was the latter set of interest politics which were to the fore – for example, in the debate surrounding the launch of 'Opportunity 2000' and in the introduction of tax relief on workplace nurseries in the budget statement of 1990.

As Ramazanoglu (1989) argues, however, feminist politics cannot afford to abandon a 'sense of sisterhood'. 'If we lose our universal generalisations, because women's oppression is only in part and in contradictory ways a universal phenomenon, then we lose the political focus of feminism and no clear feminist political strategy can be specified' (1989: 170). Somehow, out of this politics of fragmentation, women's contradictory interests have to be brought together and connected (and see too, Phillips, 1987; Segal, 1987). Constructed in and through class divisions between women, the fragmentary politics which suffuse the nanny form of waged domestic labour in contemporary Britain offer little hope in this direction. But, as we suggest now, this is not necessarily so. Although characterised by major contradictions, the possibility of political alliances between nannies and their employers does exist. And certain of these potential alliances at least are genuinely feminist strategies.

Our first cut into thinking about possible feminist politics of connection in relation to the nanny form of waged domestic labour is a position with which we are now deeply dissatisfied. However, we feel it important to articulate these arguments here, for not only were these points expressed in a variety of forms by many of the female employers whom we interviewed, but we too found ourselves sucked in by these arguments, at least initially.[3]

As befits a form of labour which for most of the post-war period in Britain has been performed by women in an unwaged capacity (for love), the employment of waged domestic labour in relation to childcare is something which is characterised by complex feelings of guilt for some, if not all, women. Indeed, with childcare constructed as home-centred care by a child's natural mother, the employment of waged domestic labour in relation to childcare in Britain has inevitably been equated with substitute mothering. In Chapter Six we showed how such circumstances structure the social relations of nanny employment in terms of a web of false kinship relations involving the female partner within a nanny-employing household, the children, the nanny and 'her' family. There is little doubt that these false kinship relations serve to mask the structural inequalities of nanny employment. Moreover, such relations also lie behind the ties of double dependency articulated by many of the middle-class mothers with whom we spoke: the social relations of this particular form of waged domestic labour therefore mean that it is not just a nanny on whom many individual middle-class women in full-time service class employment consider themselves to depend, but *a particular nanny*. Such circumstances – of caring about a particular carer – conceivably could be considered to constitute the basis for a politics of alliance between nannies and their employers. And certainly, many employers, rather than expressing a desire to challenge the reassertion of the nanny form of waged domestic labour, were keen to see improvements in nanny employment itself, and specifically in the lot of 'their' nanny. Thus, alongside suggestions such as the introduction of regulatory bodies and practices and nanny registration schemes were others geared towards improving the pay of nannies (through subsidising parental childcare costs).

In the early stages of our analysis we found such ideas attractive. Not only were they being suggested by those involved directly in nanny employment, but such ideas also constituted positive ways of countering the chief examples of abuse and bad practice which we were then meeting. Now, at some remove from the fieldwork phase of work, we are less than convinced by the long-term merit of such suggestions. For one, such policies leave unchallenged the existence of waged *domestic* labour in relation to childcare, and the traditional form of the gender division of labour on which this is based. They are, therefore, inherently traditional policies regarding childcare. Secondly, political alliances of this form between nannies and their employers ignore their grounding in the relations of *false* kinship. Such alliances would therefore be based on relations which serve to legitimate both this form of waged domestic labour and the highly traditional ideas about childcare which this reflects. Instead, we prefer to reject such possibilities in favour of others which challenge the reassertion of the nanny form of waged domestic labour directly.

There are two areas of policy which would challenge the resurgence of the nanny form of waged domestic labour. One is in the sphere of employment. The other relates to childcare service provision. As is being acknowledged increasingly within the literature on the gendering of the service class, women's paths of career advancement within such occupations are very different to those of men (see Chapter Four). Not only are women disproportionately concentrated at lower grades within service class occupations but, if they are to advance, it helps if they can be gendered as 'honorary men' (Crompton and Le Feuvre, 1992). Even then, there is considerable debate as to whether new forms of patriarchal culture are emerging so as to exclude women from career advancement (Halford and Savage, forthcoming). Although the domestic sphere remains offstage in most of these accounts, these authors are clearly fully aware of the extent to which the onset of childcare responsibilities in particular affect women in service class occupations. Effectively, such responsibilities are shown to enable employers to gender women as 'women', rather than as 'honorary men', and to see them as unsuitable candidates for promotion. Such circumstances have led Crompton, for example, to suggest the development of 'niche jobs' for women (ie., jobs which enable women in service class occupations to 'mother' if they so wish). But such suggestions merely reproduce those constructs which define women as 'homemakers' and as primary childcarers, and would re-create for the middle-class the double-day/double identity of those women working in part-time employment within dual-earner households. More preferable have to be those options which would challenge the masculine career structures which shape service class employment. Rather than constructing forms of employment which enable women to mother, one of the feminist solutions to gender inequalities within service class occupations must surely be to restructure service class employment such that it enables *both* men and women to care for their children on whatever basis they wish. Extended periods of parental leave (in contrast to the current restricted periods of maternity/paternity leave), career breaks and flexible working practices are all strategies which could be employed within service class occupations. However, to be successful they would need to be construed not as secondary niche options but as 'normal', accepted working practices by *both* men and women; a situation which would require considerable change in ideologies of work and family and their mediation by gender.

Beyond fundamental change in the nature of service class employment, the other policy changes which could challenge the resurgence of the nanny form of waged domestic labour relate to the provision of childcare. As we showed in Chapter Five, the ready supply of young, newly qualified childcare workers willing to work as nannies in the homes of the middle-classes is itself a reflection of state policy towards childcare. Such women now have little employment choice within the childcare industry beyond nannying. Moreover, it is only through the expansion of good, quality day care that childcare workers can hope to gain access to career-structured occupations with recognised, regulated terms and conditions of employment. Quality, widely available and affordable day-care, for long one of the political demands of second-wave feminism was initially construed as vital to enabling women to sell their labour power on the same terms as men. In contemporary Britain, the resurgence of waged domestic labour has shown this to be no necessity: some women can sell their labour power on the same basis as men, provided someone else (almost inevitably another woman) is 'at home' performing the housewife/mother role. Nowadays, then, campaigning for day-care has come to be as much about campaigning for 'real jobs' for those who wish to work with children, as about campaigning 'for mothers' or 'for children'. We are confident that its availability would diminish considerably the numbers of women willing to work as nannies (Chapter Five). Furthermore, we suspect that its availability would produce a crisis in labour supply as significant in its way as the 'servant crisis' of the late nineteenth/early twentieth century in Britain. Quite whether the widespread availability of good quality, affordable day-care services would reduce middle-class demand for nannies however is another matter, especially in the light of the ideological convictions underlying preferences for children to receive individual care in their own homes.

Evidently, then, there are policies which could challenge the resurgence of the nanny form of waged domestic labour, rather than reproduce it in a regulated way. Such policies too are inherently feminist. However, for them to be actively promoted requires that nannies and their employers campaign collectively for changes in childcare service provision *and* that middle-class men and women challenge the gendering of service class occupations.

In contrast to nannying, the politics of employment in home cleaning were a long way from visible during the interview stages of our research. It is not difficult to see why. By nature informal and invisible, as well as a means of supplementing income from other sources, private domestic cleaning is unlikely, in current circumstances, to become a focus for political campaigning in Britain. In Britain, then, much as Enloe has remarked in relation to the international flows of domestic labour, household cleaning ('cleaning bathrooms in the homes of the bankers') represents a *personal* response by women to reduced household incomes and the poverty consequent upon long-term reliance on state benefit (Chapter Five). In such circumstances there is no way beyond the market through which cleaning (and specifically the paltry hourly rates which the majority of working-class women earn for this) can be regulated. However, this apparently gloomy prognosis also contains possibilities. As was shown in Chapter Five, it is – for many women – the reality of life on benefit, and the structure of the benefit system in particular which produces and reproduces the need to supplement benefit-dependent incomes, and necessitates that this supplementary income be earned through the informal sector. It is, therefore, the structure of the benefit system – and the limitations which this imposes on many women's earnings – which represents the key to the politics of this form of waged domestic labour in contemporary Britain. Remove the constraints on formal sector earnings for women with partners in receipt of benefit and (notwithstanding the existence of false kinship relations within household cleaning) many of the women whom we interviewed would revert to formal sector working. Long one of the main targets of feminist political criticism, the benefit system, once more, is revealed to be a major force in the production and perpetuation of gender inequalities, but this time between women, as well as between the genders. Its reform continues to be a feminist political strategy of the first order.

Despite initial appearances to the contrary, and notwithstanding a theoretical analysis which points to the importance of divisions between women in the production and reproduction of the waged domestic labour phenomenon in contemporary Britain, there is in all this, then, a politics which can be genuinely feminist. In a world seemingly dominated by a plethora of post-modernist voices, it is a relief to us to be able to argue with conviction that fragmentary, interest-based, politics need not constitute the basis of the politics surrounding waged domestic labour. Strategies of alliance and connection between women could bring about the interim regulation of waged domestic labour, as well as its long-term elimination. Whether they will emerge, however, is the key unanswered question. Here we are forced to return to the politically negative undercurrents revealed by our analysis.

To be effective the strategies outlined above require a reassertion of those feminist arguments which recognise explicitly that it is the refusal of men to play an equal part in *all* forms of domestic labour which is one of the key means through which gender inequalities are reproduced. Somewhat lost in recent times, the force of these arguments has been well demonstrated in this volume, particularly with respect to the cleaning form of waged domestic labour, where we have seen that it is men's refusal to clean (or their inability to 'see' household dirt) which all too frequently lies behind the move to employ a cleaner. Furthermore, we have seen too how men benefit from such employment: whilst their partners occasionally continue to perform cleaning tasks (and 'manage' these workers), the men in these households reduce their participation in household cleaning to negligible levels. And the situation has been shown to be little different in the context of nanny employment. Thus, the management of nannies was revealed to be almost exclusively the domain of female partners; nannies and the female partner perform the basic work of childcare, leaving men to participate in its more pleasurable aspects. Fundamentally, these two forms of waged domestic labour represent no more than a restructuring of the traditional form of the gender division of labour. For sure, they are a means of resolving the crisis in social reproduction within individual middle-class households, but the means of resolution is one which all too clearly continues to let men 'off the hook' with respect to domestic labour. In our view it is absolutely imperative that the political responses to waged domestic labour and its resurgence in contemporary Britain recognise the above point. We would like to be confident and assert the likelihood of the recognition of the gender politics of waged domestic labour in Britain. However, we continue to have our doubts over this. Much as in an earlier epoch in Britain, the majority of those women and men whom we interviewed did not have 'a problem' with waged domestic labour *per se*. Rather, it was us (and a minority of our respondents) who had difficulties with it. It seems, then, that many dual-career households in Britain *are* willing to construct solutions to the crisis in social reproduction which are grounded in the reproduction of gender inequalities around domestic labour. It seems, too, that many of the women with whom we spoke were reluctant to acknowledge the way in which waged domestic labour works to the ultimate benefit of men. Rather, they preferred to couch things in terms of enabling *themselves* to cope with both full-time service class employment and the social reproduction of the household. Such representations we regard as profoundly depressing. Indeed, they suggest to us that ultimately many of the middle-class women who currently employ waged domestic labour are reluctant to confront – or, indeed, even see – the gender politics of domestic labour in its waged forms. Seen thus, there is, unfortunately, every likelihood that waged domestic labour will become a key feature of the mode of social reproduction within middle-class households in contemporary Britain, and that the relations of gender inequality constructed through domestic labour will, once more, become entwined with divisions of class and identity amongst women.

APPENDICES

APPENDIX 2.1 OCCUPATIONAL CATEGORIES AND TOTALS, BRITAIN: 1981–91

A Categories

1 Au pair
2 Au pair/carer
3 Au pair/driver
4 Au pair/father's help
5 Au pair/housekeeper
6 Au pair/mother's help
7 Au pair/nanny
8 Au pair/nanny/mother's help
9 Butler
10 Butler/chauffeur
11 Butler/chauffeur/cook
12 Butler/chauffeur/handyman
13 Butler/cook
14 Butler/driver
15 Butler/gardener/handyman
16 Butler/handyman
17 Carer
18 Carer/chauffeur
19 Carer/cleaner
20 Carer/cleaner/housekeeper
21 Carer/cook
22 Carer/cook/housekeeper
23 Carer/driver
24 Carer/driver/housekeeper
25 Carer/gardener
26 Carer/housekeeper
27 Carer/maternity nurse
28 Carer/mother's help
29 Carer/mother's help/nanny
30 Carer/nanny
31 Chauffeur
32 Chauffeur/cook
33 Chauffeur/cook/housekeeper

34 Chauffeur/driver
35 Chauffeur/gardener
36 Chauffeur/gardener/handyman
37 Chauffeur/handyman
38 Chauffeur/housekeeper
39 Chauffeur/mother's help
40 Cleaner
41 Cleaner/housekeeper
42 Cleaner/maid
43 Cleaner/mother's help
44 Cleaner/nanny
45 Cook
46 Cook/driver
47 Cook/driver/housekeeper
48 Cook/driver/housekeeper/nanny
49 Cook/driver/mother's help
50 Cook/driver/nanny/mother's help
51 Cook/father's help
52 Cook/gardener/housekeeper
53 Cook/housekeeper
54 Cook/housekeeper/mother's help
55 Cook/housekeeper/nanny
56 Cook/housekeeper/nanny/
   mother's help
57 Cook/handyman
58 Cook/mother's help
59 Cook/nanny/mother's help
60 Couple
61 Couple/gardener/housekeeper
62 Driver
63 Dirver/gardener
64 Driver/gardener/handyman
65 Driver/handyman

66	Driver/housekeeper	84	Groom/mother's help
67	Driver/housekeeper/ mother's help	85	Groom/nanny
		86	Handyman
68	Driver/mother's help	87	Housekeeper
69	Father's help	88	Housekeeper/handyman
70	Father's help/housekeeper	89	Housekeeper/maid
71	Father's help/housekeeper/ nanny	90	Housekeeper/mother's help
		91	Housekeeper/nanny
72	Father's help/nanny	92	Housekeeper/nanny/ mother's help
73	Footman		
74	Gardener	93	Maid
75	Gardener/handyman	94	Maid/mother's help
76	Gardener/housekeeper	95	Maternity nurse
77	Gardener/mother's help	96	Maternity nurse/nanny
78	Girl Friday	97	Mother's help
79	Girl Friday/mother's help	98	Mother's help/maternity nurse
80	Girl Friday/nanny	99	Nanny
81	Governess	100	Nanny/mother's help
82	Governess/housekeeper	101	Nanny share
83	Governess/nanny		

B Ranked categories

	Advertisements	
Occupation	*Nos*	*Percentage*
Nanny	4,692	27.1
Mother's help	2,582	14.9
Nanny/mother's help	2,392	13.8
Housekeeper	2,100	12.1
Couples	1,069	6.2
Cook/housekeeper	790	4.6
Carer/housekeeper	586	3.4
Au pair	466	2.7
Housekeeper/nanny	413	2.4
Carer	324	1.9
Au pair/mother's help	290	1.7
Housekeeper/mother's help	288	1.7
Cook	198	1.1
Maternity nurse	184	1.1
Gardener	145	0.8
Father's help	80	0.4
Girl Friday	58	0.3
Gardener/handyman	52	0.3
Butler	44	0.3
Nanny share	43	0.2
Carer/cook/housekeeper	34	0.2
Father's help/housekeeper	34	0.2
Maternity nurse/nanny	33	0.2
Au pair/nanny	29	0.2

Occupation	Nos	Percentage
Maid	28	0.2
Chauffeur	27	0.2
Carer/cook	26	0.2
Cleaner	25	0.1
Father's help/nanny	22	0.1
Handyman	21	0.1
Carer/mother's help	15	0.09
Cook/mother's help	15	0.09
Cleaner/housekeeper	13	0.08
Girl Friday/mother's help	13	0.08
Chauffeur/handyman	11	0.06
Carer/nanny	9	0.05
Girl Friday/nanny	9	0.05
Driver/housekeeper	8	0.05
Governess	8	0.05
Butler/cook	7	0.04
Driver	7	0.04
Gardener/housekeeper	7	0.04
Governess/nanny	7	0.04
Au pair/housekeeper	6	0.04
Au pair/nanny/mother's help	6	0.04
Carer/driver/housekeeper	6	0.04
Chauffeur/gardener	6	0.04
Cleaner/mother's help	6	0.04
Cook/driver/housekeeper	6	0.04
Housekeeper/maid	6	0.04
Butler/handyman	<5	0.03
Cook/driver		
Cook/housekeeper/mother's help		
Cook/housekeeper/nanny		
Driver/gardener/handyman		
Driver/handyman		
Father's help/housekeeper/nanny		
Au pair/carer		
Butler/chauffeur		
Housekeeper/nanny/mother's help		
Butler/driver		
Carer/driver		
Carer/maternity nurse		
Chauffeur/cook		
Chauffeur/mother's help		
Cook/nanny/mother's help		
Driver/gardener		
Governess/housekeeper		
Groom/nanny		
Mothers help/maternity nurse		
Au pair/driver		
Au pair/father's help		
Butler/chauffeur/cook		
Butler/chauffeur/handyman		

Occuption	Nos	Percentage
Butler/gardener/handyman		
Carer/chauffeur		
Carer/cleaner		
Carer/cleaner/housekeeper		
Carer/gardener		
Carer/mother's help/nanny		
Chauffeur/cook/housekeeper		
Chauffeur/driver		
Chauffeur/housekeeper		
Cleaner/maid		
Cleaner/nanny		
Cook/driver/housekeeper/nanny		
Cook/driver/mother's help		
Cook/driver/nanny/mother's help		
Cook/father's help		
Cook/gardener/housekeeper		
Cook/housekeeper/nanny/mother's help		
Cook/handyman		
Couple/gardener/housekeeper		
Driver/housekeeper/mother's help		
Driver/mother's help		
Footman		
Gardener/mother's help		
Groom/mother's help		
Housekeeper/handyman		
Maid/mother's help		
Total	17,334	

APPENDIX 2.2 OCCUPATIONAL CATEGORIES AND TOTALS, NORTH-EAST AND SOUTH-EAST ENGLAND: 1981–91

A North-east England

Category	July–Dec. 81	1982	1983	1984	1985	1986	1987	1988	1989	1990	Jan.–July 1991	Total
Nanny	3 (10.3%)	6 (17.1%)	12 (21.0%)	15 (22.0%)	21 (35.6%)	24 (34.8%)	14 (24.6%)	38 (38.8%)	55 (47.8%)	45 (45.9%)	13 (40.6%)	**246 (34.3%)**
Nanny/ Mother's Help	6 (20.7%)	3 (8.6%)	1 (1.7%)	8 (11.8%)	8 (13.5%)	9 (13.0%)	6 (10.5%)	11 (11.2%)	21 (18.3%)	16 (16.3%)	5 (15.6%)	**94 (13.1%)**
Mother's Help	4 (13.8%)	6 (17.1%)	15 (26.3%)	20 (29.4%)	7 (11.9%)	13 (18.8%)	7 (12.3%)	27 (27.5%)	18 (15.6%)	9 (9.2%)	2 (6.2%)	**128 (17.8%)**
Cleaner	12 (41.4%)	16 (45.7%)	20 (35.1%)	17 (25.0%)	19 (32.2%)	14 (20.3%)	21 (36.8%)	18 (18.4%)	11 (9.6%)	20 (20.4%)	8 (25.0%)	**176 (24.5%)**
Housekeeper	4 (13.8%)	4 (11.4%)	9 (15.8%)	7 (10.3%)	4 (6.8%)	9 (13.0%)	9 (15.8%)	4 (4.1%)	10 (8.7%)	8 (8.2%)	4 (12.5%)	**72 (10.0%)**
Au pair	–	–	–	1 (1.5%)	–	–	–	–	–	–	–	**1 (0.1%)**
Total private adverts	29	35	57	68	59	69	57	98	115	98	32	**(717)**

Source: Newcastle Evening Chronicle

APPENDIX 2.2 (CONT.) OCCUPATIONAL CATEGORIES AND TOTALS NORTH-EAST AND SOUTH-EAST ENGLAND: 1981–91

B South-east England

Category	July–Dec. 81	1982	1983	1984	1985	1986	1987	1988	1989	1990	Jan.–July 1991	Total
Cleaner	3 (6.5%)	6 (10.3%)	4 (6.2%)	6 (6.8%)	24 (18.8%)	13 (10.0%)	14 (12.1%)	21 (16.2%)	20 (18.2%)	19 (21.8%)	7 (26.9%)	137 (13.9%)
Mother's help	7 (15.2%)	13 (22.4%)	17 (26.1%)	35 (39.8%)	42 (32.8%)	44 (33.8%)	34 (29.3%)	29 (22.5%)	20 (18.2%)	16 (18.3%)	8 (30.7%)	265 (27.0%)
Nanny/ Mother's help	0 –	3 (5.2%)	5 (7.7%)	10 (11.4%)	8 (6.2%)	11 (8.4%)	12 (10.3%)	4 (3.1%)	13 (11.8%)	3 (3.4%)	1 (3.8%)	70 (7.1%)
Au pair/ Mother's help	0 –	0 –	0 –	0 –	0 –	1 (0.8%)	4 (3.4%)	4 (3.1%)	0 –	0 –	0 –	9 (0.9%)
Au pair	2 (4.3%)	0 –	0 –	0 –	2 (1.5%)	0 –	4 (3.4%)	2 (1.5%)	1 (0.9%)	1 (1.1%)	0 –	12 (1.2%)
Nanny	0 –	5 (8.6%)	8 (12.3%)	10 (11.4%)	20 (15.6%)	15 (11.5%)	13 (11.2%)	20 (15.5%)	22 (20.0%)	23 (26.4%)	3 (11.5%)	139 (14.1%)
Nanny-share	0 –	0 –	0 –	0 –	0 –	0 –	2 (1.7%)	0 –	2 (1.8%)	1 (1.1%)	2 (7.7%)	7 (0.7%)
Housekeeper	11 (23.9%)	11 (19.0%)	18 (27.7%)	14 (15.9%)	10 (7.8%)	10 (7.7%)	12 (10.3%)	20 (15.5%)	13 (11.8%)	11 (12.6%)	0 –	130 (13.2%)
After school carer	0 –	0 –	0 –	0 –	3 (2.3%)	6 (4.6%)	3 (2.6%)	9 (6.9%)	8 (7.2%)	0 –	0 –	29 (3.0%)
Couple	7 (15.2%)	4 (6.9%)	3 (4.6%)	1 (1.1%)	2 (1.5%)	5 (3.8%)	3 (2.6%)	1 (0.8%)	1 (0.9%)	2 (2.3%)	0 –	29 (3.0%)

APPENDIX 2.2 (CONT.)

B South-east England (cont.)

Category	July-Dec. 81	1982	1983	1984	1985	1986	1987	1988	1989	1990	Jan.-July 1991	Total
Cook	5 (10.9%)	3 (5.2%)	1 (1.5%)	0 –	1 (0.8%)	0 –	0 –	0 –	0 –	2 (2.3%)	0 –	12 (1.2%)
Carer	4 (8.6%)	1 (1.7%)	1 (1.5%)	2 (2.3%)	4 (3.1%)	11 (8.5%)	3 (2.6%)	7 (5.4%)	1 (0.9%)	4 (4.6%)	0 –	38 (3.9%)
Maternity Nurse	0 –	0 –	0 –	0 –	1 (0.8%)	0 –	0 –	0 –	0 –	0 –	0 –	1 (10.1%)
Girl Friday	1 (2.2%)	0 –	0 –	0 –	0 –	0 –	0 –	0 –	0 –	0 –	0 –	1 (0.1%)
Handyman	0 –	1 (1.7%)	0 –	0 –	2 (1.6%)	0 –	0 –	0 –	0 –	0 –	0 –	3 (0.3%)
Driver/ Handyman	1 (2.2%)	0 –	0 –	0 –	0 –	0 –	0 –	0 –	0 –	0 –	0 –	1 (0.1%)
Gardener/ Handyman	2 (4.3%)	0 –	2 (3.0%)	0 –	3 (2.3%)	2 (1.5%)	0 –	0 –	3 (2.7%)	0 –	0 –	12 (1.2%)
Gardener	2 (4.3%)	10 (17.2%)	6 (9.2%)	10 (11.4%)	6 (4.6%)	12 (9.2%)	11 (9.5%)	12 (9.3%)	6 (5.4%)	5 (5.7%)	5 (19.2%)	85 (8.6%)
Chauffeur	1 (2.2%)	1 (1.7%)	0 –	0 –	0 –	0 –	1 (0.9%)	0 –	0 –	0 –	0 –	3 (0.3%)
Total private adverts	46	58	65	88	128	130	116	129	110	87	26	983

Source: Reading Chronicle

APPENDIX 2.3: OCCUPATIONAL CATEGORIES BY COUNTY, ENGLAND, 1981–91

A County ranking, England: 1981

Rank	County	Recorded advertisements	Estimated total	Percentage
1	Surrey	2011	16,088	25.4
2	Berkshire	649	5,192	8.2
3	Kent	605	4,840	7.6
4	Hertfordshire	517	4,136	6.5
5	Hampshire	484	3,872	6.1
6	West Sussex	419	3,352	5.3
7	Buckinghamshire	416	3,328	5.3
8	Oxfordshire	320	2,560	4.0
9	Essex	270	2,160	3.4
10	Gloucestershire	265	2,120	3.3
11	Wiltshire	202	1,616	2.6
12	Cambridgeshire	185	1,480	2.3
13	Devon	158	1,264	1.9
14	Avon	126	1,008	1.5
=15	Cheshire	120	960	1.5
	East Sussex	120	960	1.5
16	Northamptonshire	118	944	1.5
17	Dorset	99	792	1.3
18	Suffolk	93	744	1.2
19	Hereford/Worcestershire	87	696	1.1
=20	Cornwall	80	640	1.0
	Somerset	80	640	1.0
=21	North Yorkshire	73	584	0.9
	Warwickshire	73	584	0.9
22	West Midlands	70	560	0.9
23	Leicestershire	66	528	0.8
=24	Norfolk	63	504	0.8
	West Yorkshire	63	504	0.8
25	Bedfordshire	57	456	0.7
26	Lincolnshire	53	424	0.7
27	Lancashire	52	416	0.7
28	Shropshire	41	328	0.5
29	Greater Manchester	39	312	0.5
=30	Cumbria	37	296	0.5
	Derbyshire	37	296	0.5
31	Nottinghamshire	32	256	0.4
=32	Staffordshire	28	224	0.4
	Northumberland	28	224	0.4
33	Isle of Wight	26	208	0.3
34	Cleveland	25	200	0.3
35	South Yorkshire	22	176	0.3
36	Merseyside	20	160	0.3
37	Durham	16	128	0.2
38	Tyne & Wear	8	64	0.1
39	Humberside	7	56	0.1

APPENDIX 2.3 (CONT.)

B location quotients for English counties and metropolitan areas, excluding London

1 All advertisements

Rank	County	Percentage of advertised total*	Percentage of total households**	LQ
1	Surrey	25.4	2.3	11.0
2	Berkshire	8.2	1.5	5.5
3	Buckinghamshire	5.3	1.3	4.1
4	Oxfordshire	4.0	1.1	3.6
5	West Sussex	5.3	1.6	3.3
6	Hertfordshire	6.5	2.1	3.1
7	Gloucestershire	3.3	1.1	3.0
8	Kent	7.6	3.3	2.3
9	Cambridgeshire	2.3	1.1	2.1
10	Hampshire	6.1	3.5	1.7
11	Wiltshire	2.6	1.6	1.6
12	Northamptonshire	1.5	1.2	1.3
=13	Isle of Wight	0.3	0.3	1.0
	Essex	3.4	3.3	1.0
	Cornwall	1.0	1.0	1.0
	Somerset	1.0	1.0	1.0
=14	Dorset	1.3	1.4	0.9
	East Sussex	1.5	1.7	0.9
	Devon	1.9	2.2	0.9
	Suffolk	1.2	1.4	0.9
=15	Warwickshire	0.9	1.1	0.8
	Hereford/Worcestershire	1.1	1.4	0.8
=16	Cheshire	1.5	2.1	0.7
	Avon	1.4	2.1	0.7
=17	Bedfordshire	0.7	1.1	0.6
	Shropshire	0.5	0.8	0.6
	North Yorkshire	0.9	1.5	0.6
	Northumberland	0.4	0.7	0.6
=18	Norfolk	0.8	1.6	0.5
	Lincolnshire	0.7	1.3	0.5
=19	Leicestershire	0.8	1.9	0.4
	Cumbria	0.5	1.1	0.4
20	Cleveland	0.3	1.2	0.3
=21	Derbyshire	0.5	2.1	0.2
	Lancashire	0.7	3.2	0.2
	Nottinghamshire	0.4	2.3	0.2
	West Yorkshire	0.8	4.8	0.2
	Staffordshire	0.4	2.3	0.2
=22	Durham	0.2	1.4	0.1
	South Yorkshire	0.3	3.0	0.1
	Greater Manchester	0.5	6.0	0.1
	West Midlands	0.9	5.9	0.1
	Merseyside	0.3	3.4	0.1
	Humberside	0.1	1.9	0.1
=23	Tyne & Wear	0.1	2.7	0.04

Notes: * Appendix 2.3 A
      ** 1981 Census

APPENDIX 2.3 B (CONT.)

2 Nanny advertisements

Rank	County	Percentage of nanny advertisements*	Percentage of households with 1 person aged 0–4**	LQ
1	Surrey	25.2	1.9	13.3
2	Berkshire	7.7	1.5	5.1
3	Oxfordshire	5.0	1.1	4.5
4	Hertfordshire	7.8	1.8	4.3
5	Buckinghamshire	5.0	1.3	3.8
6	Gloucestershire	2.9	1.0	2.9
7	West Sussex	3.4	1.3	2.6
8	Kent	7.1	3.1	2.3
9	Cambridgeshire	2.7	1.3	2.1
10	Hampshire	5.5	3.2	1.7
=11	Isle of Wight	0.3	0.2	1.5
	Wiltshire	1.6	1.1	1.5
12	Suffolk	1.6	1.3	1.2
13	Cornwall	1.0	0.8	1.2
14	Avon	1.9	1.8	1.1
=15	Northamptonshire	1.2	1.2	1.0
	Somerset	0.8	0.8	1.0
	Dorset	0.9	1.0	1.0
=16	Warwickshire	0.8	1.0	0.8
	Cheshire	1.5	2.0	0.8
17	Norfolk	1.0	1.4	0.7
18	Bedfordshire	0.8	1.3	0.6
=19	Devon	0.9	1.8	0.5
	Lincoln	0.6	1.2	0.5
	North Yorkshire	0.6	1.3	0.5
=20	Northumberland	0.3	0.7	0.4
	Essex	1.4	3.2	0.4
	Hereford/Worcestershire	0.6	1.4	0.4
	Cleveland	0.5	1.4	0.4
=21	Leicestershire	0.6	1.9	0.3
	Lancashire	0.8	2.9	0.3
	Shropshire	0.2	0.8	0.3
	Nottinghamshire	0.6	2.2	0.3
	Derbyshire	0.5	2.0	0.3
=22	East Sussex	0.6	3.2	0.2
	West Midlands	1.0	5.5	0.2
	West Yorkshire	0.8	4.4	0.2
=23	Cumbria	0.1	1.0	0.1
	South Yorkshire	0.3	2.8	0.1
	Staffordshire	0.3	2.5	0.1
	Durham	0.1	1.5	0.1
	Merseyside	0.2	3.2	0.1
	Greater Manchester	0.3	5.9	0.1
24	Tyne & Wear	0.1	2.7	0.04

Notes: * Appendix 2.3 A (no advertisements were recorded for Humberside)
      ** 1981 Census

APPENDIX 2.3 B (CONT.)

3 Housekeeper advertisements

Rank	County	Percentage of housekeeper* advertisements	Percentage of total** households	LQ
1	Surrey	14.4	2.3	6.3
2	West Sussex	7.1	1.6	4.4
3	Buckinghamshire	5.5	1.3	4.2
= 4	Berkshire	5.9	1.5	3.9
	Gloucestershire	4.3	1.1	3.9
5	Oxfordshire	3.7	1.1	3.4
6	Hertfordshire	5.1	2.1	2.4
= 7	Kent	7.1	3.3	2.2
	Cambridgeshire	2.4	1.1	2.2
= 8	Hampshire	5.5	3.5	1.6
	Northamptonshire	1.9	1.2	1.6
	Somerset	1.6	1.0	1.6
9	Devon	3.2	2.2	1.5
10	Wiltshire	2.2	1.6	1.4
=11	East Sussex	2.0	1.7	1.2
	Avon	2.6	2.1	1.2
=12	Essex	3.5	3.3	1.1
	North Yorkshire	1.7	1.5	1.1
	Dorset	1.6	1.4	1.1
	Hereford/Worcestershire	1.6	1.4	1.1
	Lincolnshire	1.4	1.3	1.1
	Shropshire	0.9	0.8	1.1
13	Northumberland	0.7	0.7	1.0
=14	Cheshire	1.9	2.1	0.9
	Cornwall	0.9	1.0	0.9
=15	Warwickshire	0.8	1.1	0.7
	Suffolk	0.9	1.4	0.7
16	Norfolk	0.9	1.6	0.6
17	Leicestershire	0.9	1.9	0.5
=18	Bedfordshire	0.4	1.1	0.4
	Cumbria	0.4	1.1	0.4
=19	Isle of Wight	0.1	0.3	0.3
	Lancashire	0.9	3.2	0.3
	Staffordshire	0.8	2.3	0.3
	Derbyshire	0.6	2.1	0.3
=20	West Midlands	1.3	5.9	0.2
	West Yorkshire	0.9	4.8	0.2
	Nottinghamshire	0.5	2.3	0.2
	Cleveland	0.3	1.2	0.2
=21	South Yorkshire	0.3	3.0	0.1
	Durham	0.2	1.4	0.1
	Merseyside	0.2	3.4	0.1
	Tyne & Wear	0.2	2.7	0.1
	Humberside	0.1	1.9	0.1

Not present: Greater Manchester.

Notes: * Appendix 2.3 A
 ** 1981 Census

APPENDIX 2.4 CATEGORIES OF ADVERTISED DEMAND BY COUNTY, ENGLAND, 1981–91

1 NANNY (n = 1493)

		Total recorded advertisements	Percentage
1	Surrey	367	25.2
2	Hertfordshire	113	7.8
3	Berkshire	112	7.7
4	Kent	104	7.1
5	Hampshire	80	5.5
= 6	Oxfordshire	73	5.0
	Buckinghamshire	73	5.0
7	West Sussex	50	3.4
8	Gloucestershire	42	2.9
9	Cambridgeshire	39	2.7
10	Avon	28	1.9
=11	Suffolk	24	1.6
	Wiltshire	24	1.6
12	Cheshire	23	1.5
13	Essex	21	1.4
14	Northamptonshire	17	1.2
=15	Norfolk	15	1.0
	West Midlands	15	1.0
16	Cornwall	14	1.0
=17	Devon	13	0.9
	Dorset	13	0.9
=18	Bedfordshire	12	0.8
	Warwickshire	12	0.8
	Lancashire	12	0.8
=19	Somerset	11	0.7
	West Yorkshire	11	0.7
=20	East Sussex	9	0.6
	Hereford/Worcestershire	9	0.6
	Leicestershire	9	0.6
=21	Nottinghamshire	8	0.5
	Lincolnshire	8	0.5
	North Yorkshire	8	0.5
	Cleveland	7	0.5
	Derbyshire	7	0.5
=22	Isle of Wight	5	0.3
	Northumberland	5	0.3
	South Yorkshire	5	0.3
	Staffordshire	4	0.3
	Greater Manchester	4	0.3
=23	Merseyside	3	0.2
	Shropshire	3	0.2
=24	Cumbria	2	0.1
	Durham	2	0.1
	Tyne & Wear	2	0.1

Not present: Humberside.

2 HOUSEKEEPER (n = **1462**)

		Total recorded advertisements	Percentage
1	Surrey	200	14.4
2	Kent	100	7.1
3	West Sussex	98	7.1
4	Berkshire	82	5.9
= 5	Hampshire	77	5.5
	Buckinghamshire	77	5.5
6	Hertfordshire	71	5.1
7	Gloucestershire	60	4.3
8	Oxfordshire	52	3.7
9	Essex	48	3.5
10	Devon	44	3.2
11	Avon	36	2.6
12	Cambridgeshire	33	2.4
13	Wiltshire	30	2.2
14	East Sussex	28	2.0
=15	Cheshire	27	1.9
16	Northamptonshire	26	1.9
17	North Yorkshire	24	1.7
18	Dorset	22	1.6
=19	Hereford/Worcestershire	21	1.6
	Somerset	21	1.6
20	Lincolnshire	19	1.4
21	West Midlands	18	1.3
=22	Cornwall	13	0.9
	Lancashire	13	0.9
	Leicestershire	13	0.9
	Norfolk	13	0.9
	Shropshire	13	0.9
	Suffolk	13	0.9
	West Yorkshire	12	0.9
=23	Staffordshire	11	0.8
	Warwickshire	11	0.8
24	Northumberland	9	0.7
25	Derbyshire	8	0.6
=26	Greater Manchester	7	0.5
	Nottinghamshire	7	0.5
=27	Bedfordshire	6	0.4
	Cumbria	6	0.4
=28	Cleveland	4	0.3
	South Yorkshire	4	0.3
=29	Durham	3	0.2
	Merseyside	3	0.2
	Tyne & Wear	3	0.2
=30	Isle of Wight	2	0.1
	Humberside	2	0.1

All counties present.

3 MOTHER'S HELP (n = 926)

		Total recorded advertisements	Percentage
1	Surrey	233	26.0
2	Berkshire	88	9.8
3	Hertfordshire	77	8.6
4	Kent	69	7.1
5	Buckinghamshire	49	5.5
6	Hampshire	46	5.1
7	Oxfordshire	42	4.7
8	West Sussex	32	3.6
9	Essex	26	2.9
10	Devon	21	2.3
11	Wiltshire	19	2.1
12	Gloucestershire	17	1.9
13	Cambridgeshire	16	1.8
14	Avon	13	1.4
15	Northamptonshire	11	1.2
=16	Cheshire	10	1.1
	East Sussex	10	1.1
	Suffolk	10	1.1
	Hereford/Worcestershire	10	1.1
=17	Bedfordshire	8	0.9
	North Yorkshire	8	0.9
=18	Norfolk	7	0.8
	West Midlands	7	0.8
=19	Dorset	6	0.7
	Greater Manchester	6	0.7
	Warwickshire	6	0.7
=20	Cornwall	5	0.6
	Somerset	5	0.6
=21	Cleveland	4	0.5
	Cumbria	4	0.5
	Isle of Wight	4	0.5
	Lincolnshire	4	0.5
=22	Lancashire	3	0.3
	Leicestershire	3	0.3
	Northumberland	3	0.3
	Shropshire	3	0.3
	West Yorkshire	3	0.3
=23	Derbyshire	2	0.2
	Staffordshire	2	0.2
=24	Durham	1	0.1
	Nottinghamshire	1	0.1
	South Yorkshire	1	0.1

Not present: Humberside; Merseyside; Tyne & Wear.

4 NANNY/MOTHER'S HELP (n = 835)

		Total recorded advertisements	Percentage
1	Surrey	232	27.7
2	Berkshire	80	9.6
3	Hertfordshire	72	8.6
4	Kent	66	7.8
5	Buckinghamshire	50	5.9
6	West Sussex	36	4.3
7	Hampshire	34	4.1
8	Essex	30	3.6
9	Oxfordshire	24	2.9
10	Gloucestershire	23	2.7
11	Cambridgeshire	15	1.8
12	Cheshire	14	1.7
13	Northamptonshire	13	1.5
14	Wiltshire	12	1.4
=15	Devon	10	1.2
	Warwickshire	10	1.2
16	Somerset	9	1.1
=17	Avon	8	1.0
	East Sussex	8	1.0
	Hereford/Worcestershire	8	1.0
	South Yorkshire	8	1.0
18	West Yorkshire	7	0.8
=19	Dorset	5	0.6
	Leicestershire	5	0.6
	North Yorkshire	5	0.6
=20	Bedfordshire	4	0.5
	Norfolk	4	0.5
=21	Cornwall	3	0.4
	Isle of Wight	3	0.4
	West Midlands	3	0.4
=22	Greater Manchester	2	0.2
	Lancashire	2	0.2
	Lincolnshire	2	0.2
	Nottinghamshire	2	0.2
	Shropshire	2	0.2
=23	Derbyshire	1	0.1
	Merseyside	1	0.1
	Northumberland	1	0.1

Not present: Cleveland; Cumbria; Durham; Humberside; Staffordshire; Suffolk; Tyne & Wear.

5 COUPLES (n = 617)

		Total recorded advertisements	Percentage
1	Surrey	132	15.1
2	Berkshire	90	10.3
3	Hampshire	69	7.9
4	Kent	65	7.4
5	Buckinghamshire	62	7.1
6	Gloucestershire	49	5.6
7	West Sussex	47	5.9
8	Wiltshire	41	4.7
9	Hertfordshire	35	4.0
10	Essex	32	3.7
11	Oxfordshire	28	3.2
12	East Sussex	17	1.9
13	Northamptonshire	16	1.8
=14	Devon	15	1.7
	Dorset	15	1.7
=15	Cambridgeshire	14	1.6
	Suffolk	14	1.6
16	Hereford/Worcestershire	13	1.5
=17	Avon	10	1.1
	Leicestershire	10	1.1
=18	Cornwall	8	0.9
	Cumbria	8	0.9
	Lancashire	8	0.9
19	Lincolnshire	7	0.8
=20	Somerset	6	0.7
	West Midlands	6	0.7
=21	Cheshire	5	0.6
	Greater Manchester	5	0.6
	Norfolk	5	0.6
	North Yorkshire	5	0.6
	Shropshire	5	0.6
	Warwickshire	5	0.6
=22	Cornwall	4	0.5
	Northumberland	4	0.5
	South Yorkshire	4	0.5
23	Bedfordshire	3	0.4
=24	Cleveland	2	0.2
	Isle of Wight	2	0.2
	Nottinghamshire	2	0.2
	Tyne & Wear	2	0.2
	West Yorkshire	2	0.2
25	Staffordshire	1	0.1

Not present: Durham; Humberside; Merseyside.

6 COOK/HOUSEKEEPER (n = 534)

		Total recorded advertisements	Percentage
1	Surrey	67	13.2
2	Kent	44	8.6
3	Berkshire	38	7.4
4	Hampshire	37	7.2
5	West Sussex	33	6.5
6	Buckinghamshire	23	4.5
7	Oxfordshire	21	4.1
= 8	Essex	20	3.9
	Hertfordshire	20	3.9
	Gloucestershire	20	3.9
9	Wiltshire	18	3.5
10	Cambridgeshire	16	3.1
11	Devon	12	2.4
=12	East Sussex	11	2.2
	Northamptonshire	9	1.8
=13	Somerset	8	1.6
	Hereford/Worcestershire	8	1.6
=14	Dorset	7	1.4
	North Yorkshire	7	1.4
	Warwickshire	7	1.4
	West Midlands	7	1.4
=15	Cornwall	6	1.2
	Leicestershire	6	1.2
	Suffolk	6	1.2
=16	Bedfordshire	5	1.0
	Isle of Wight	5	1.0
	Cheshire	5	1.0
	Lincolnshire	5	1.0
=17	Lancashire	4	0.8
	Norfolk	4	0.8
	Shropshire	4	0.8
=18	Avon	3	0.6
	Derbyshire	3	0.6
	Greater Manchester	3	0.6
	Merseyside	3	0.6
	Northumberland	3	0.6
	West Yorkshire	3	0.6
=19	Cleveland	2	0.4
	Durham	2	0.4
	Nottinghamshire	2	0.4
	Staffordshire	2	0.4
20	South Yorkshire	1	0.2

Not present: Cumbria; Humberside; Tyne & Wear.

7   CARER/HOUSEKEEPER (n = 429)

		Total recorded advertisements	Percentage
1	Surrey	52	12.7
2	Berkshire	34	8.3
3	Hampshire	30	7.6
4	Hertfordshire	29	7.4
5	Kent	28	6.9
6	West Sussex	26	6.4
7	Devon	20	4.9
8	Essex	18	4.4
9	Buckinghamshire	17	4.2
10	Wiltshire	15	3.7
=11	Cornwall	11	2.7
	Gloucestershire	11	2.7
	Oxfordshire	11	2.7
12	Cambridgeshire	9	2.2
13	Cheshire	8	1.9
=14	East Sussex	7	1.7
	Hereford/Worcestershire	7	1.7
	West Yorkshire	7	1.7
=15	North Yorkshire	6	1.5
	Suffolk	6	1.5
	West Midlands	6	1.5
16	Leicestershire	5	1.2
=17	Avon	4	1.0
	Somerset	4	1.0
	Warwickshire	4	1.0
=18	Dorset	3	0.7
	Cumbria	3	0.7
	Lincolnshire	3	0.7
	Norfolk	3	0.7
=19	Bedfordshire	2	0.5
	Derbyshire	2	0.5
	Durham	2	0.5
	Northamptonshire	2	0.5
	Nottinghamshire	2	0.5
	Shropshire	2	0.5
	South Yorkshire	2	0.5
=20	Humberside	1	0.3
	Lancashire	1	0.3
	Merseyside	1	0.3
	Northumberland	1	0.3
	Staffordshire	1	0.3
	Tyne & Wear	1	0.3

Not present: Cleveland; Isle of Wight; Greater Manchester.

8 HOUSEKEEPER/NANNY **(n = 226)**

		Total recorded advertisements	Percentage
1	Surrey	46	20.4
2	Kent	25	11.1
3	Berkshire	21	9.3
4	Hertfordshire	17	7.5
= 5	Hampshire	12	5.3
	Oxfordshire	12	5.3
6	West Sussex	11	4.9
7	Buckinghamshire	8	3.5
= 8	Cambridgeshire	7	3.1
	Essex	7	3.1
= 9	Cheshire	5	2.2
	Dorset	5	2.2
	Gloucestershire	5	2.2
	Northamptonshire	5	2.2
10	East Sussex	4	1.8
=11	Derbyshire	3	1.3
	Lancashire	3	1.3
=12	Avon	2	0.9
	Bedfordshire	2	0.9
	Cornwall	2	0.9
	Cumbria	2	0.9
	Devon	2	0.9
	Greater Manchester	2	0.9
	Hereford/Worcestershire	2	0.9
	Staffordshire	2	0.9
	Suffolk	2	0.9
	Warwickshire	2	0.9
	Wiltshire	2	0.9
=13	Cleveland	1	0.4
	Leicestershire	1	0.4
	Nottinghamshire	1	0.4
	Shropshire	1	0.4
	South Yorkshire	1	0.4
	West Midlands	1	0.4
	West Yorkshire	1	0.4

Not present: Durham; Humberside; Isle of Wight; Lincolnshire; Merseyside; Norfolk; North Yorkshire; Northumberland; Somerset; Tyne & Wear.

9 CARER (n = 207)

		Total recorded advertisements	Percentage
1	Surrey	22	10.6
2	Kent	14	6.8
= 3	Berkshire	13	6.3
	West Sussex	13	6.3
4	Hampshire	12	5.8
= 5	Hertfordshire	11	5.3
	Oxfordshire	11	5.3
6	Wiltshire	8	3.9
= 7	Cambridgeshire	7	3.4
	Dorset	7	3.4
	West Yorkshire	7	3.4
8	Essex	6	2.9
= 9	Buckinghamshire	5	2.4
	Cornwall	5	2.4
	East Sussex	5	2.4
	Northamptonshire	5	2.4
=10	Bedfordshire	4	1.9
	Leicestershire	4	1.9
	Norfolk	4	1.9
	Nottinghamshire	4	1.9
=11	Cheshire	3	1.4
	Devon	3	1.4
	Gloucestershire	3	1.4
	Suffolk	3	1.4
	Warwickshire	3	1.4
=12	Avon	2	1.0
	Cumbria	2	1.0
	Greater Manchester	2	1.0
	Hereford/Worcestershire	2	1.0
	Lancashire	2	1.0
	Somerset	2	1.0
	South Yorkshire	2	1.0
=13	Cleveland	1	0.5
	Durham	1	0.5
	Humberside	1	0.5
	Lincolnshire	1	0.5
	Merseyside	1	0.5
	Staffordshire	1	0.5
	Tyne & Wear	1	0.5
	West Midlands	1	0.5

Not present: Isle of Wight; North Yorkshire; Northumberland; Shropshire; Derbyshire.

10 COOK (n = 154)

		Total recorded advertisements	Percentage
1	Hampshire	17	11.3
2	Surrey	15	9.7
3	Berkshire	12	7.8
= 4	West Sussex	10	6.5
	Wiltshire	10	6.5
5	Oxfordshire	8	5.2
= 6	Buckinghamshire	7	4.5
	Gloucestershire	7	4.5
	Kent	7	4.5
7	North Yorkshire	6	3.9
8	Cambridgeshire	5	3.3
= 9	Essex	4	2.6
	Hereford/Worcestershire	4	2.6
	Hertfordshire	4	2.6
=10	Avon	3	1.9
	Cheshire	3	1.9
	Dorset	3	1.9
	East Sussex	3	1.9
	Norfolk	3	1.9
	West Midlands	3	1.9
	West Yorkshire	3	1.9
=11	Bedfordshire	2	1.3
	Cumbria	2	1.3
	Leicestershire	2	1.3
	Lincolnshire	2	1.3
	Northumberland	2	1.3
	Somerset	2	1.3
	Suffolk	2	1.3
=12	Cornwall	1	0.7
	Devon	1	0.7
	Greater Manchester	1	0.7

Not present: Cleveland; Derbyshire; Durham; Humberside; Isle of Wight; Lancashire; Merseyside; Northamptonshire; Nottinghamshire; Shropshire; South Yorkshire; Staffordshire; Tyne & Wear; Warwickshire.

11 GARDENER **(n = 136)**

		Total recorded advertisements	Percentage
1	Surrey	20	14.7
2	Berkshire	16	11.8
3	Hampshire	15	11.0
4	Kent	12	8.8
5	Buckinghamshire	11	8.1
= 6	Oxfordshire	8	5.9
	West Sussex	8	5.9
7	Gloucestershire	7	5.1
= 8	Hertfordshire	5	3.7
	Wiltshire	5	3.7
= 9	Cambridgeshire	3	2.2
	East Sussex	3	2.2
	Warwickshire	3	2.2
=10	Avon	2	1.5
	Essex	2	1.5
	North Yorkshire	2	1.5
	Northamptonshire	2	1.5
	Shropshire	2	1.5
	Somerset	2	1.5
	West Yorkshire	2	1.5
=11	Bedfordshire	1	0.7
	Cheshire	1	0.7
	Isle of Wight	1	0.7
	Lancashire	1	0.7
	Leicestershire	1	0.7
	Nottinghamshire	1	0.7

Not present: Cleveland; Cornwall; Cumbria; Derbyshire; Devon; Dorset; Durham; Greater Manchester; Hereford and Worcestershire; Humberside; Lincolnshire; Merseyside; Norfolk; Northumberland; South Yorkshire; Staffordshire; Suffolk; Tyne & Wear; West Midlands.

12 AU PAIR **(n = 132)**

		Total recorded advertisements	Percentage
1	Surrey	41	31.1
2	Kent	13	9.8
= 3	Buckinghamshire	8	6.1
	Hertfordshire	8	6.1
4	West Sussex	7	5.3
= 5	Avon	5	3.8
	Berkshire	5	3.8
= 6	Devon	4	3.0
	Gloucestershire	4	3.0
= 7	Bedfordshire	3	2.3
	Cleveland	3	2.3
	Essex	3	2.3
	Hampshire	3	2.3
	Shropshire	3	2.3
	Suffolk	3	2.3
= 8	Cambridgeshire	2	1.5
	Isle of Wight	2	1.5
	Leicestershire	2	1.5
= 9	Cheshire	1	0.8
	Dorset	1	0.8
	Durham	1	0.8
	East Sussex	1	0.8
	Hereford/Worcestershire	1	0.8
	Lancashire	1	0.8
	Merseyside	1	0.8
	Oxfordshire	1	0.8
	Warwickshire	1	0.8
	West Midlands	1	0.8

Not present: Cornwall; Cumbria; Derbyshire; Humberside; Lincolnshire; Norfolk; North Yorkshire; Northamptonshire; Northumberland; Nottinghamshire; Somerset; South Yorkshire; Staffordshire; Tyne & Wear; West Yorkshire; Wiltshire.

13 HOUSEKEEPER/MOTHER'S HELP (n = 131)

		Total recorded advertisements	Percentage
1	Surrey	31	23.7
2	Hertfordshire	14	10.7
3	Berkshire	12	9.2
4	Kent	9	6.9
5	West Sussex	8	6.1
6	Hampshire	7	5.3
7	Essex	5	3.8
= 8	Avon	4	3.1
	Buckinghamshire	4	3.1
	Cheshire	4	3.1
	Gloucestershire	4	3.1
	Oxfordshire	4	3.1
9	Somerset	3	2.3
=10	Cambridgeshire	2	1.5
	Devon	2	1.5
	East Sussex	2	1.5
	Greater Manchester	2	1.5
	Norfolk	2	1.5
	Nottinghamshire	2	1.5
=11	Cumbria	1	0.8
	Dorset	1	0.8
	Hereford/Worcestershire	1	0.8
	Leicestershire	1	0.8
	Northamptonshire	1	0.8
	Shropshire	1	0.8
	Warwickshire	1	0.8
	West Yorkshire	1	0.8
	Wiltshire	1	0.8

Not present: Bedfordshire; Cleveland; Cornwall; Derbyshire; Durham; Humberside; Isle of Wight; Lancashire; Lincolnshire; Merseyside; North Yorkshire; Northumberland; South Yorkshire; Staffordshire; Suffolk; Tyne & Wear; West Midlands.

14 AU PAIR/MOTHER'S HELP (n = 82)

		Total recorded advertisements	Percentage
1	Surrey	24	29.3
2	Hertfordshire	8	9.7
3	Kent	7	8.5
4	Berkshire	6	7.3
5	Hampshire	5	6.1
=6	Oxfordshire	4	4.9
	West Sussex	4	4.9
7	Buckinghamshire	3	3.6
=8	Avon	2	2.4
	Cambridgeshire	2	2.4
	Cornwall	2	2.4
	Norfolk	2	2.4
=9	Bedfordshire	1	1.2
	Cheshire	1	1.2
	Cleveland	1	1.2
	Devon	1	1.2
	Durham	1	1.2
	East Sussex	1	1.2
	Essex	1	1.2
	Gloucestershire	1	1.2
	North Yorkshire	1	1.2
	Northamptonshire	1	1.2
	Somerset	1	1.2
	Staffordshire	1	1.2
	Wiltshire	1	1.2

15 GARDENER/HANDYMAN (n = 50)

		Total recorded advertisements	Percentage
1	West Sussex	8	16.0
2	Surrey	7	14.0
3	Hampshire	5	10.0
=4	Essex	3	6.0
	Hertfordshire	3	6.0
	Kent	3	6.0
	Wiltshire	3	6.0
=5	Berkshire	2	4.0
	Buckinghamshire	2	4.0
	East Sussex	2	4.0
	Oxfordshire	2	4.0
	Warwickshire	2	4.0
=6	Cheshire	1	2.0
	Cumbria	1	2.0
	Dorset	1	2.0
	Leicestershire	1	2.0
	Northamptonshire	1	2.0
	Somerset	1	2.0
	Suffolk	1	2.0
	West Midlands	1	2.0

16 FATHER'S HELP (**n** = **49**)

		Total recorded advertisements	Percentage
1	Surrey	10	Not calculated
2	Hampshire	7	from category
3	Essex	6	16 owing to
4	Kent	4	small numbers
=5	Buckinghamshire	3	recorded
	Hertfordshire	3	
	West Sussex	3	
=6	Bedfordshire	2	
	Cambridgeshire	2	
=7	Avon, Berkshire; Cornwall;		
	East Sussex; Gloucestershire;		
	Leicestershire; Northamptonshire;		
	Somerset; Wiltshire	1	

17 GIRL FRIDAY (**n** = **35**)

		Total recorded advertisements
1	Berkshire	5
2	Oxfordshire	4
=3	Devon	3
	Gloucestershire	3
=4	Buckinghamshire	2
	Cambridgeshire	2
	Dorset	2
	Hampshire	2
	Suffolk	2
	Surrey	2
=5	Avon; Cornwall; Derbyshire; East Sussex;	
	Hertfordshire; Kent; Lancashire; West Sussex	1

18 MATERNITY NURSE (**n** = **32**)

		Total recorded advertisements
1	Surrey	5
2	Hertfordshire	4
=3	Berkshire	3
	Buckinghamshire	3
	Kent	3
	Wiltshire	3
4	Hampshire	2
=5	Bedfordshire; Cambridgeshire; Devon; East	
	Sussex; Essex; Oxfordshire; Warwickshire;	
	West Sussex; West Yorkshire	1

19 BUTLER **(n = 31)**

		Total recorded advertisements
=1	Buckinghamshire	4
	Hertfordshire	4
=2	Cambridgeshire	3
	West Sussex	3
=3	Berkshire	2
	Cumbria	2
	Wiltshire	2
=4	Dorset; Essex; Gloucestershire; Hampshire; Humberside; Kent; Lincolnshire; Northamptonshire; Oxfordshire; Suffolk; Surrey	1

20 HOUSEKEEPER/FATHER'S HELP **(n = 24)**

		Total recorded advertisements
1	Surrey	5
2	Kent	3
=3	Cheshire	2
	East Sussex	2
	Essex	2
	Hertfordshire	2
	Oxfordshire	2
=4	Buckinghamshire; Devon; Hampshire; Northamptonshire; Somerset; Warwickshire	1

21 CARER/COOK/HOUSEKEEPER **(n = 23)**

		Total recorded advertisements
1	Surrey	6
=2	Cambridgeshire	2
	Norfolk	2
	Kent	2
=3	Avon; Bedfordshire; Cornwall; Devon; Durham; Essex; Gloucestershire; Northamptonshire; Oxfordshire; Staffordshire; West Yorkshire	1

22 CARER/COOK **(n = 26)**

		Total recorded advertisements
1	Surrey	5
2	Hertfordshire	3
=3	Bedfordshire	2
	Buckinghamshire; Devon; Dorset; Norfolk; Oxfordshire; Somerset	

23 FATHER'S HELP/NANNY (n = 16)

		Total recorded advertisements
1	Surrey	4
=2	Bedfordshire	2
	Dorset	2
	Kent	2
=3	Hertfordshire; Cambridgeshire; Northamptonshire; Warwickshire; West Sussex; Wiltshire	1

24 HANDYMAN (n = 15)

		Total recorded advertisements
1	West Sussex	3
=2	Buckinghamshire	2
	Dorset	2
	East Sussex	2
	Hampshire	2
3	Cornwall; Cumbria; Oxfordshire; Surrey	1

25 CLEANER (n = 13)

		Total recorded advertisements
1	Surrey	3
2	Buckinghamshire	2
3	Cornwall; Devon; Dorset; East Sussex; Hampshire; Kent; West Sussex	1

26 MATERNITY NURSE/NANNY (n = 11)

		Total recorded advertisements
=1	Buckinghamshire; Cambridgeshire; Gloucestershire; Greater Manchester; Kent; Norfolk; Northamptonshire; Surrey; Warwickshire; West Sussex; Wiltshire	1

27 CHAUFFEUR (n = 9)

		Total recorded advertisements
1	Kent	3
=2	Staffordshire	2
	Berkshire	2
3	West Sussex; West Yorkshire	1

CHAUFFEUR/HANDYMAN (n = 9)

		Total recorded advertisements
1	Kent	3
2	Leicestershire	2
=3	East Sussex; Hampshire; Lincolnshire; Shropshire; Somerset; Surrey	1

COOK/MOTHER'S HELP (n = 9)

		Total recorded advertisements
1	Hertfordshire	3
=2	Oxfordshire	2
	Surrey	2
3	Devon; Kent	1

MAID (n = 9)

		Total recorded advertisements
1	Hertfordshire	5
2	Berkshire; Cheshire; Surrey; West Yorkshire	1

28 AU PAIR/NANNY (n = 7)

		Total recorded advertisements
1	Surrey	4
2	Buckinghamshire; Hertfordshire; Somerset	1

DRIVER/HOUSEKEEPER (n = 7)

		Total recorded advertisements
1	Kent	2
=2	Hampshire; Norfolk; Northamptonshire; Surrey; West Sussex	1

GARDENER/HOUSEKEEPER (n = 7)

		Total recorded advertisements
=1	Wiltshire	2
	Berkshire; Cornwall; Dorset; Hampshire; Surrey	1

CLEANER/HOUSEKEEPER (n = 7)

		Total recorded advertisements
=1	Cumbria; Hampshire; Hertfordshire; Lancashire; Oxfordshire; Surrey; West Sussex	1

27 COOK/DRIVER/HOUSEKEEPER (n = 6)

		Total recorded advertisements
=1	Buckinghamshire	2
	Surrey	2
2	Hertfordshire; Oxfordshire	1

GIRL FRIDAY/MOTHER'S HELP (n = 6)

		Total recorded advertisements
1	West Sussex	2
=2	Buckinghamshire; Hampshire; Shropshire; Surrey	1

NANNY SHARE (n = 6)

		Total recorded advertisements
1	Surrey	4
2	Berkshire	2

28 CHAUFFEUR/GARDENER (n = 5)

		Total recorded advertisements
=1	Kent	2
	Wiltshire	2
2	Oxfordshire	1

DRIVER/GARDENER/HANDYMAN (n = 5)

		Total recorded advertisements
1	Gloucestershire	2
2	Hampshire; Suffolk; West Sussex	1

29 CARER/DRIVER/HOUSEKEEPER (n = 4)

		Total recorded advertisements
1	Norfolk	2
2	Hertfordshire; West Sussex	1

CHAUFFEUR/GARDENER/HANDYMAN (n = 4)

		Total recorded advertisements
1	Dorset; Kent; Oxfordshire; Surrey	1

DRIVER (n = 4)

		Total recorded advertisements
=1	Buckinghamshire; Hampshire; Surrey; West Midlands	1

FATHER'S HELP/HOUSEKEEPER/NANNY (n = 4)

		Total recorded advertisements
1	Durham; Essex; Hertfordshire; Surrey	1

30 AU PAIR/NANNY/MOTHER'S HELP (n = 3)

		Total recorded advertisements
1	Surrey	2
2	Northamptonshire	1

BUTLER/COOK (n = 3)

Total recorded advertisements

1   Berkshire   2
2   Cambridgeshire   1

CARER/NANNY (n = 3)

Total recorded advertisements

1   Hampshire; Kent; Surrey   1

COOK/HOUSEKEEPER/MOTHER'S HELP (n = 3)

Total recorded advertisements

1   Cheshire; Cumbria; Kent   1

GOVERNESS (n = 3)

Total recorded advertisements

1   Cambridgeshire; Cornwall; Norfolk   1

GOVERNESS/NANNY (n = 3)

Total recorded advertisements

1   Buckinghamshire; Hampshire; Surrey   1

31 AU PAIR/HOUSEKEEPER (n = 2)

Total recorded advertisements

1   Hereford/Worcestershire; Surrey   1

CHAUFFEUR/MOTHER'S HELP (n = 2)

Total recorded advertisements

1   Leicestershire   2

DRIVER/GARDENER (n = 2)

Total recorded advertisements

1   Kent   2

GROOM/NANNY (n = 2)

Total recorded advertisements

1   Durham; Warwickshire   1

HOUSEKEEPER/MAID (n = 2)

Total recorded advertisements

1   Berkshire; Surrey   1

HOUSEKEEPER/NANNY/MOTHER'S HELP (n = 2)

1   Berkshire; Essex   1

(n = 1)

AU PAIR/DRIVER	Warwickshire
AU PAIR/FATHER'S HELP	Dorset
BUTLER/CHAUFFEUR	Hertfordshire
BUTLER/CHAUFFEUR/COOK	Kent
BUTLER/CHAUFFEUR/HANDYMAN	Dorset
BUTLER/GARDENER/HANDYMAN	Surrey
BUTLER/HANDYMAN	West Sussex
CARER/CLEANER/HOUSEKEEPER	Derbyshire
CARER/DRIVER	Buckinghamshire
CARER/GARDENER	Devon
CARER/MOTHER'S HELP/NANNY	Cambridgeshire
CHAUFFEUR/COOK	Cheshire
CHAUFFEUR/COOK/HOUSEKEEPER	Cornwall
COOK/HOUSEKEEPER/NANNY	Hertfordshire
COOK/HOUSEKEEPER/NANNY/MOTHER'S HELP	West Sussex
COOK/HANDYMAN	Oxfordshire
COOK/MOTHER'S HELP/NANNY	Kent
DRIVER/HANDYMAN	Surrey
DRIVER/MOTHER'S HELP	West Sussex
GARDENER/CHAUFFEUR	Cumbria
GARDENER/MOTHER'S HELP	West Sussex
GIRL FRIDAY/NANNY	Surrey
GOVERNESS/HOUSEKEEPER	Cornwall
GROOM/MOTHER'S HELP	Cornwall
HOUSEKEEPER/HANDYMAN	Cheshire
MAID/MOTHER'S HELP	Kent
COOK/GARDENER/HOUSEKEEPER	Essex

APPENDIX 2.5 CATEGORIES OF ADVERTISED DEMAND BY POSTCODE AREA, LONDON, 1981–91

1 NANNY (n = 3,199)

Postcode area		Numbers	Percentage
1	South-West	1,239	38.7
2	North-West	736	23.0
3	West	514	16.1
4	North	440	13.8
5	South-East	172	5.4
6	West-Central	55	1.7
7	East	39	1.2
8	East-Central	4	0.1

All postcode areas present

2 MOTHER'S HELP (n = 1,656)

Postcode area		Numbers	Percentage
1	South-West	608	36.7
2	North-West	555	33.5
3	North	213	12.9
4	West	188	11.4
5	South-East	54	3.3
6	West-Central	24	1.5
7	East	11	0.7
8	East-Central	3	0.2

All postcode areas present

3 NANNY/MOTHER'S HELP (n = 1,557)

Postcode area		Numbers	Percentage
1	South-West	620	39.8
2	North-West	450	28.9
3	West	201	12.9
4	North	187	12.0
5	South-East	69	4.4
6	West-Central	17	1.1
7	East	10	0.6
8	East-Central	3	0.2

All postcode areas present

4 HOUSEKEEPER (n = 638)

Postcode area		Numbers	Percentage
1	North-West	255	40.0
2	South-West	168	26.3
3	West	117	18.3
4	North	72	11.3
=5	South-East	11	1.7
	West-Central	11	1.7
=6	East	2	0.3
	East-Central	2	0.3

All postcode areas present

5 AU PAIR (n = 327)

Postcode area	Numbers	Percentage
1 North-West	118	36.1
2 South-West	102	31.2
3 West	49	15.0
4 North	40	12.2
=5 South-East	7	2.1
West-Central	7	2.1
6 East	3	0.9
7 East-Central	1	0.3

All postcode areas present

6 COOK/HOUSEKEEPER (n = 254)

Postcode area	Numbers	Percentage
1 North-West	94	37.0
2 South-West	57	22.4
3 West	51	20.1
4 North	35	13.8
5 West-Central	12	4.7
6 South-East	3	1.2
=7 East	1	0.4
East-Central	1	0.4

All postcode areas present

7 AU PAIR/MOTHER'S HELP (n = 201)

Postcode area	Numbers	Percentage
1 North-West	83	41.3
2 South-West	70	34.8
3 West	23	11.4
4 North	17	8.5
5 South-East	6	3.0
6 West-Central	2	1.0

Not present: East; East-Central

8 HOUSEKEEPER/NANNY (n = 178)

Postcode area	Numbers	Percentage
1 South-West	71	39.9
2 North-West	48	27.0
3 West	31	17.4
4 North	19	10.7
=5 South-East	4	2.3
West-Central	4	2.3
6 East	1	0.6

Not present: East-Central

9 HOUSEKEEPER/MOTHER'S HELP (n = 153)

Postcode area		Numbers	Percentage
1	North-West	57	37.3
2	South-West	48	31.4
3	North	23	15.0
4	West	18	11.8
5	West-Central	3	2.0
6	South-East	2	1.3
=7	East	1	0.7
	East-Central	1	0.7

All postcode areas present

10 MATERNITY NURSE (n = 151)

Postcode area		Numbers	Percentage
1	South-West	71	46.4
2	North-West	49	32.0
3	West	19	12.4
4	North	10	6.5
=5	South-East	1	0.7
	West-Central	1	0.7

Not present: East; East-Central

11 CARER/HOUSEKEEPER (n = 150)

Postcode area		Numbers	Percentage
1	North-West	60	40.0
2	South-West	30	20.0
3	West	26	17.3
4	North	22	14.7
5	South-East	5	3.3
6	West-Central	4	2.7
7	East	2	1.3
8	East-Central	1	0.7

All postcode areas present

12 COUPLE (n = 132)

Postcode area		Numbers	Percentage
1	South-West	41	31.1
2	West	37	28.0
3	North-West	33	25.0
4	North	8	6.1
5	West-Central	7	5.3
=6	East-Central	3	2.3
	South-East	3	2.3

Not present: East

13 CARER (n = 109)

Postcode area		Numbers	Percentage
1	North-West	35	32.1
2	South-West	29	26.6
3	West	19	17.4
4	North	14	12.8
5	West-Central	6	5.5
6	South-East	5	4.6
7	East	1	0.9

Not present: East-Central

14 NANNY SHARE (n = 37)

Postcode area		Numbers	Percentage
1	South-West	16	43.2
2	North	10	27.0
3	South-East	4	10.8
=4	West	3	8.1
	North-West	3	8.1
5	East-Central	1	2.7

Not present: East; West-Central

15 COOK (n = 33)

Postcode area		Numbers	Percentage
1	South-West	14	42.4
2	North-West	9	27.3
3	West	8	24.2
=4	North	1	3.0
	West-Central	1	3.0

Not present: East; East-Central; South-East

16 FATHER'S HELP (n = 31)

Postcode area		Numbers	Percentage
1	North-West	16	51.6
2	South-West	7	22.6
3	West	4	12.9
4	North	3	9.7
5	West-Central	1	3.2

Not present: East; East-Central; South East

17 MATERNITY NURSE/NANNY (n = 22)

Postcode area		Numbers	Percentage
1	North-West	8	36.4
2	South-West	7	31.8
3	West	4	18.2
=4	North	1	4.6
	East-Central	1	4.6
	South-East	1	4.6

Not present: East; West-Central

18 AU PAIR/NANNY (n = 21)

Postcode area		Numbers	Percentage
1	South-West	9	42.9
2	North	5	23.8
3	West-Central	3	14.3
4	West	2	9.5
5	East	1	4.8
	North-West	1	4.8

Not present: East-Central; South-East

19 CHAUFFEUR (n = 18)

Postcode area		Numbers	Percentage
1	West	6	33.3
2	South-West	5	27.8
3	North-West	4	22.2
4	North	2	11.1
5	South-East	1	5.6

Not present: East; East-Central; West-Central

20 MAID (n = 18)

Postcode area		Numbers	Percentage
=1	West	6	33.3
	South-West	6	33.3
2	North-West	4	22.2
3	North	1	5.6
	West-Central	1	5.6

Not present: East; East-Central; South-East

21 GIRL FRIDAY (n = 17)

Postcode area		Numbers	Percentage
1	South-West	12	70.6
2	West	3	17.6
3	North-West	2	11.8

Not present: North; East; East-Central; South-East; West-Central

22 BUTLER (n = 11)

Postcode area		Numbers	Percentage
1	West	6	54.6
2	South-West	4	36.4
3	North-West	1	9.1

Not present: North; East; East-Central; South-East; West-Central

23 CARER/COOK (n = 10)

Postcode area		Numbers	Percentage
1	South-West	5	50.0
2	West	3	30.0
3	North	2	20.0

Not present: East; North-West; East-Central; South-East; West-Central

24 CARER/COOK/HOUSEKEEPER (n = 10)

Postcode area	Numbers	Percentage
1 North-West	5	50.0
2 West	2	20.0
=3 North	1	10.0
East-Central	1	10.0
South-West	1	10.0

Not present: East; South-East; West-Central

25 CLEANER (n = 10)

Postcode area	Numbers	Percentage
1 North-West	4	40.0
2 South-West	3	30.0
3 North	2	20.0
4 West-Central	1	10.0

Not present: East; West; East-Central; South-East

26 FATHER'S HELP/HOUSEKEEPER (n = 9)

Postcode area	Numbers	Percentage
1 North-West	4	44.4
=2 West	2	22.2
South-West	2	22.2
3 North	1	11.1

Not present: East; North-West; East-Central; South-East; West-Central

27 GIRL FRIDAY/NANNY (n = 8)

Postcode area	Numbers	Percentage
1 North-West	3	37.5
=2 West	2	25.0
South-West	2	25.0
3 North	1	12.5

Not present: East; East-Central; South-East; West-Central

28 CARER/NANNY (n = 6)

Postcode area	Numbers	Percentage
1 South-West	3	50.0
2 West	2	33.3
3 North	1	16.7

Not present: East; North-West; East-Central; South-East; West-Central

29 COOK/MOTHER'S HELP (n = 6)

Postcode area	Numbers	Percentage
1 North-West	4	66.7
2 West	2	33.3

Not present: North; East; East-Central; South-East; South-West; West-Central

30 CLEANER/HOUSEKEEPER (n = 6)

Postcode area	Numbers	Percentage
1 West	2	33.3
=2 North	1	16.7
North-West	1	16.7
South-West	1	16.7
West-Central	1	16.7

Not present: East; East-Central; South-East.

31 CLEANER/MOTHER'S HELP (n = 6)

Postcode area	Numbers	Percentage
1 North-West	6	100.00

Not present: North; East; West; East-Central; South-West; West-Central

32 FATHER'S HELP/NANNY (n = 6)

Postcode area	Numbers	Percentage
1 South-West	5	83.3
2 West	1	16.7

Not present: North; East; North-West; East-Central; South-East; West-Central

33 GIRL FRIDAY/MOTHER'S HELP (n = 6)

Postcode area	Numbers	Percentage
1 North-West	4	66.7
=2 North	1	16.7
West	1	16.7

Not present: East; East-Central; South-East; South-West; West-Central

34 GARDENER (n = 5)

Postcode area	Numbers	Percentage
1 North-West	3	60.0
2 South-East	2	40.0

Not present: North; East; West; East-Central; South-West; West-Central

35 GOVERNESS (n = 5)

Postcode area	Numbers	Percentage
=1 West	2	40.0
South-West	2	40.0
2 North-West	1	20.0

Not present: North; East; East-Central; South-East; West-Central

36 HANDYMAN (n = 5)

Postcode area	Numbers	Percentage
=1 North	2	40.0
North-West	2	40.0
2 West-Central	1	20.0

Not present: East; West; East-Central; South-East; South-West

37 BUTLER/COOK (n = 4)

Postcode area	Numbers	Percentage
1 South-West	4	100.0

Not present: North; East; West; North-West; East-Central; South-East; West-Central

38 BUTLER/HANDYMAN (n = 4)

Postcode area	Numbers	Percentage
1 South-West	2	50.0
=2 North	1	25.0
West	1	25.0

Not present: East; North-West; East-Central; South-East; West-Central

39 COOK/HOUSEKEEPER/NANNY (n = 4)

Postcode area	Numbers	Percentage
1 South-West	2	50.0
=2 West	1	25.0
North-West	1	25.0

Not present: North; East; East-Central; South-East; West-Central

40 DRIVER/HANDYMAN (n = 4)

Postcode area	Numbers	Percentage
1 North-West	4	100.0

Not present: North; East; West; East-Central; South-East; South-West; West-Central

41 GOVERNESS/NANNY (n = 4)

Postcode area	Numbers	Percentage
1 North-West	2	50.0
=2 West	1	25.0
South-East	1	25.0

Not present: North; East; East-Central; South-West; West-Central

42 AU PAIR/CARER (n = 3)

Postcode area	Numbers	Percentage
1 North-West	3	100.0

Not present: North; East; East-Central; South-East; South-West; West-Central

43 AU PAIR/HOUSEKEEPER (n = 3)

Postcode area	Numbers	Percentage
=1 West	1	33.3
North-West	1	33.3
West-Central	1	33.3

Not present: North; East; East-Central; South-East; South-West

281

44 AU PAIR/NANNY/MOTHER'S HELP (n = 3)

Postcode area	Numbers	Percentage
=1　North	1	33.3
North-West	1	33.3
South-West	1	33.3

Not present: East; West; East-Central; South-East; West-Central

45 CARER/MOTHER'S HELP (n = 3)

Postcode area	Numbers	Percentage
1　North-West	2	66.7
2　South-West	1	33.3

Not present: North; East; West; East-Central; South-East; West-Central

46 BUTLER/CHAUFFEUR (n = 2)

Postcode area	Numbers	Percentage
=1　West	1	50.0
South-West	1	50.0

Not present: North; East; North-West; East-Central; South-East; West-Central

47 BUTLER/DRIVER (n = 2)

Postcode area	Numbers	Percentage
=1　West	1	50.0
South-West	1	50.0

Not present: North; East; North-West; East-Central; South-Central; South-East; West-Central

48 CARER/DRIVER/HOUSEKEEPER (n = 2)

Postcode area	Numbers	Percentage
1　North-West	2	100.0

Not present: North; East; West; East-Central; South-East; South-West; West-Central

49 CARER/MATERNITY NURSE (n = 2)

Postcode area	Numbers	Percentage
=1　North-West	1	50.0
South-West	1	50.0

Not present: North; East; West; East-Central; South-East; West-Central

50 CHAUFFEUR/HANDYMAN (n = 2)

Postcode area	Numbers	Percentage
=1　West	1	50.0
South-West	1	50.0

Not present: North; East; North-West; East-Central; South-East; West-Central

51 COOK/DRIVER (n = 2)

Postcode area	Numbers	Percentage
=1 North	1	50.0
North-West	1	50.0

Not present: East; West; East-Central; South-East; South-West; West-Central

52 COOK/HOUSEKEEPER/MOTHER'S HELP (n = 2)

Postcode area	Numbers	Percentage
=1 West	1	50.0
South-West	1	50.0

Not present: North; East; North-West; East-Central; South-East; West-Central

53 DRIVER (n = 2)

Postcode area	Numbers	Percentage
=1 West	1	50.0
South-West	1	50.0

Not present: North; East; North-West; East-Central; South-East; West-Central

54 MATERNITY NURSE/MOTHER'S HELP (n = 2)

Postcode area	Numbers	Percentage
=1 East	1	50.0
West	1	50.0

Not present: North; North-West; East-Central; South-East; South-West; West-Central

55 (n = 1)

CARER/CLEANER	North-West
CARER/DRIVER	North-West
CHAUFFEUR/DRIVER	South-West
CHAUFFEUR/HOUSEKEEPER	North-West
COOK/DRIVER/MOTHER'S HELP	South-West
COOK/DRIVER/NANNY/MOTHER'S HELP	North-West
COOK/FATHER'S HELP	West
COOK/NANNY/MOTHER'S HELP	North-West
CLEANER/MAID	North
CLEANER/NANNY	South-West
DRIVER/GARDENER/HANDYMAN	North-West
DRIVER/HOUSEKEEPER	South-East
DRIVER/HOUSEKEEPER/MOTHER'S HELP	North
FOOTMAN	West
GARDENER/HANDYMAN	South-West
GOVERNESS/HOUSEKEEPER	North-West
HOUSEKEEPER/MOTHER'S HELP/NANNY	North-West

APPENDIX 2.6 THE NORTH-EAST AND SOUTH-EAST WORKPLACE SURVEYS

Both the north-east and south-east surveys were accessed through a range of workplaces known to employ women in social class I/II (i.e., professional/ managerial) occupations. Both surveys were conducted sequentially in the Newcastle and Reading areas, i.e., between July and October of 1991. In the north-east a total of 1,140 questionnaires were distributed; in the south-east, 977. The intention was that these 'drops' would yield approximately 300 dual-career households in both areas with both partners in full-time social class I/II employment. The workplace composition of each survey was as follows:

	North-east		South-east	
	Nos	Percentage	Nos	Percentage
A Public sector				
Health	310	27.3	311	31.8
Local Government	308	27.1	70	7.2
Education	238	21.0	252	25.8
B Private sector				
Financial services	161	14.2	159	16.3
Professional services (excluding financial)	83	7.3	137	14.0
Literary/artistic	8	0.7	23	2.4
Retail (management)	28	2.5	25	2.6

Although it would have been preferable to have sampled workplaces randomly, in practice the distribution of professionals and/or managers meant that we had to approach 'known' and/or potential employers. The survey was discussed with personnel staff and was administered internally by them (usually with pay rolls) after discussion of the grade levels pertinent to the study. The numbers of questionnaires distributed therefore reflect personnel's estimates of the extent of female professional/managerial employment within specific organisations, as well as variable levels of co-operation. They do not represent an attempt to quota sample the distribution of professional/managerial employment in the two local labour markets.

In the end, this strategy yielded 268 full-time dual-career partnerships in the north-east and 274 in the south-east. In order to make comparisons with Savage *et al.*'s work on the new middle-classes, we reclassified these occupations by social order. 253 questionnaire survey returns in the north-east fell into social orders 1–4 and 264 in the south-east. The full breakdown of this is as follows:

A *Both partners in the same*	North-east		South-east	
social order	Nos	Percentage	Nos	Percentage
1 Senior, national, local government	95	37.5	69	26.1
2 Education, welfare, health	49	19.4	67	25.3
3 Literary, artistic, sports	–		–	
4 Scientific/technical	1	0.4	8	3.0

B Both partners in different social orders	North-east		South-east	
	Nos	*Percentage*	*Nos*	*Percentage*
1 Female in social order 1				
Male in social order 2	14	5.5	7	2.7
Male in social order 3	1	0.4	1	0.4
Male in social order 4	30	11.6	17	6.4
2 Female in social order 2				
Male in social order 1	24	9.5	39	14.8
Male in social order 3	4	1.6	7	2.7
Male in social order 4	29	11.5	41	15.5
3 Female in social order 3				
Male in social order 1	1	0.4	–	–
Male in social order 2	1	0.4	–	–
Male in social order 4	–	–	–	–
4 Female in social order 4				
Male in social order 1	3	1.2	3	1.2
Male in social order 2	1	0.4	5	1.9
Male in social order 3	–	–	–	–

The major difference between the two sets of returns is one which could be anticipated from 1981 census data, i.e., the concentration of north-east returns within social orders one and two. Otherwise, the major characteristics of the two sets of returns are predictable, i.e., the concentration of female professionals in social order two and the importance of social order four for their partners.

APPENDIX 5.1 LABOUR FORCE CHARACTERISTICS

A NANNIES (n = 25 in each region)
5.1a Age

Categories	North-east		South-east	
	Nos	Percentage	Nos	Percentage
17 and under	–	–	–	–
18–20	8	40	5	21
21–23	8	32	11	46
24–26	6	24	7	28
over 27	3	12	2	8

Sources: Fieldwork interview data

5.1b Marital status

Categories	North-east		South-east	
	Nos	Percentage	Nos	Percentage
Single	21	84	18	72
Married	3	12	7	28
Separated/divorced	1	4	–	–

Sources: Fieldwork interview data

5.1c Residential details

Categories	North-east		South-east	
	Nos	Percentage	Nos	Percentage
Parental home	15	60	9	36
Married/cohabiting	7	28	11	44
'Live-in' in employers' home	1	4	5	20
Single parent househould	1	4	–	–
Single person household	1	4	–	–

Sources: Fieldwork interview data

5.1d Social class of parental household
a) North-east

Social class of father	Social class of mother						Out of labour force
	I	II	IIIN	IIIM	IV	V	
I (Professional)	–	–	1 (4%)	–	–	–	–
II (Intermediate)	–	1(4%)	6(24%)	2(8%)	1(4%)	–	2(8%)
IIIN (Skilled, non-manual)	–	–	1 (4%)	–	–	–	–
IIIM (Skilled, manual)	–	–	3(12%)	1(4%)	1(4%)	–	1(4%)
IV (Semi-skilled)	–	–	–	–	–	–	–
V (Unskilled)	–	–	–	–	–	–	–
Out of labour force/unknown	–	–	1 (4%)	–	2(8%)	–	2(8%)

286

b) South-east

Social class of father	Social class of mother						Out of labour force
	I	II	IIIN	IIIM	IV	V	
I (Professional)	2(8%)	1(4%)	–	–	–	–	3(12%)
II (Intermediate)	–	1(4%)	–	–	1 (4%)	–	1 (4%)
IIIN (Skilled, non-manual)	–	–	–	–	–	–	–
IIIM (Skilled, manual)	–	–	–	–	3(12%)	–	4(16%)
IV (Semi-skilled)	–	1(4%)	–	–	4(16%)	–	1 (4%)
V (Unskilled)	–	–	–	–	–	–	–
Out of labour force/unknown	–	–	–	–	2 (8%)	–	1 (4%)

Source: Fieldwork interview data

5.1e Educational qualifications
a) School-leaving age

Categories	North-east		South-east	
	Nos	Percentage	Nos	Percentage
16+	24	96	23	92
18+	1	4	2	8

Sources: Fieldwork interview data

b) Highest educational qualification on leaving school

Categories	North-east		South-east	
	Nos	Percentage	Nos	Percentage
CSEs	18	72	5	20
O levels	6	24	17	68
A levels	–		2	8
None	1	4	1	4

Sources: Fieldwork interview data

5.1f Childcare and related training

Categories	North-east		South-east	
	Nos	Percentage	Nos	Percentage
NNEB (i) state	12	48	12	48
(ii) private	–	–	2	8
PCSC	7	28	4	16
None	6	24	7	28

Sources: Fieldwork interview data

5.1g Length of time employed as a nanny*

Categories	North-east		South-east	
	Nos	Percentage	Nos	Percentage
<6 months	1	4	–	–
6 mths–1 year	3	12	2	8
1–2 years	3	12	3	12
2–4 years	16	64	15	60
>5 years	2	8	5	20

Sources: Fieldwork interview data
Note: * This period may include employment in > 1 job.

B CLEANERS (n = 10)
5.1h Age

Categories	North-east		South-east	
	Nos	Percentage	Nos	Percentage
<40	–	–	3	30
41–50	4	40	1	10
51–60	3	30	–	–
61–70	2	20	4	40
>70	1	10	2	20

Sources: Fieldwork interview data

5.1i Marital status

Categories	North-east		South-east	
	Nos	Percentage	Nos	Percentage
Single	–	–	–	–
Married	7	70	7	70
Separated/divorced	2	20	1	10
Widowed	1	10	2	20

Sources: Fieldwork interview data

5.1j Social class
a) Husband (by last/current occupation)

Social class	North-east		South-east	
	Nos	Percentage	Nos	Percentage
I	–	–	–	–
II	–	–	1	10
IIIN	–	–	–	–
IIIM	–	–	7	70
IV	9	90	–	–
V	1	10	2	20

Sources: Fieldwork interview data

b) Father (by last occupation)

Social class	North-east		South-east	
	Nos	Percentage	Nos	Percentage
I	—	—	1	10
II	1	10	—	—
IIIN	—	—	—	—
IIIM	—	—	5	50
IV	9	90	3	30
V	—	—	1	10

Sources: Fieldwork interview data

c) Self (by last previous occupation)

Social class	North-east		South-east	
	Nos	Percentage	Nos	Percentage
I	—	—	—	—
II	—	—	1	10
IIIN	—	—	—	—
IIIM	—	—	1	10
IV	3	30	4	40
V	7	70	4	40

Sources: Fieldwork interview data

5.1k Educational qualifications
a) School leaving age

Categories	North-east		South-east	
	Nos	Percentage	Nos	Percentage
14+	4	40	5	50
15+	6	60	2	20
16+	—	—	3	30

Sources: Fieldwork interview data

b) Highest educational qualification on leaving school

Categories	North-east		South-east	
	Nos	Percentage	Nos	Percentage
School certificate	—	—	—	—
O levels	—	—	1	10
None	10	100	9	90

Sources: Fieldwork interview data

5.11 Length of time employed as a cleaner*

Categories	North-east		South-east	
	Nos	*Percentage*	*Nos*	*Percentage*
<6 months	–	–	–	–
6 mths–1 year	–	–	2	20
1–2 years	2	20	–	–
2–4 years	3	30	1	10
>5 years	5	50	7	70

Sources: Fieldwork interview data
* This period may include employment in >1 job.

APPENDIX 6.1 NANNY EMPLOYMENT: TERMS AND CONDITIONS

6.1.1 Pay (after deductions)*

Pay bands (per week)	North-east	South-east
< £65	1	–
£66–80	13	1
£81–100	9	5
£101–120	2	9
£121–140	–	5
£141–160	–	4
> £160	–	1
	25	25

Source: Fieldwork interview data

Note *This figure is the 'take-home' pay which a nanny receives from her employer(s). In the case of nanny shares we have included all employer contributions. However, where a nanny's 'take-home' is made up from one primary employer and a number of secondary casual 'top ups', the 'top ups' have not been included.

6.1.2: Contracts

Category	North-east	South-east
Signed contract of employment	3 (12%)	20 (80%)
No contract of employment	22 (88%)	5 (20%)

Sources: Fieldwork interview data

6.1.3: Holidays*

Category	North-east	South-east
None**	2 (8%)	–
3 weeks	4 (16%)	1 (4%)
4 weeks	14 (56%)	19 (76%)
5 weeks	3 (12%)	2 (12%)
6 weeks	2 (8%)	3 (8%)

Sources: Fieldwork interview data

Notes *i.e., paid holidays, beyond public holidays. The majority of nannies are required to take at least two weeks of their annual holiday allowance when employers are 'on holiday'. Typically, therefore, they have between one and two weeks to take at their convenience.

**None = no paid holiday.

6.1.4: 'Perks' (excluding accommodation)

	Category	North-east	South-east
	Cars		
a)	Car provided by employers	–	10 (40%)
b)	Assistance towards car running costs	5 (20%)	11 (44%)
	'Holidays'*		
a)	Fully paid overseas holidays	1 (4%)	1 (4%)
b)	Fully paid UK holidays	1 (4%)	2 (8%)
	Poll Tax/Community charge**		
a)	Paid fully by employers	1 (4%)	4 (16%)
b)	Assistance by employers	–	7 (28%)

Source: Fieldwork interview data.

Notes * Although considered to be 'perks', these are not usually 'holidays' for the nanny, who remains in charge of the child(ren) but in a different work context/ environment.

** The policy context has, of course, changed since the period of our fieldwork.

Tables 6.1.1, 6.1.2. and 6.1.4. suggest some key differences *vis-à-vis* the terms and conditions of nanny employment in the north-east and south-east. However, given the small size of our interview sample, it is impossible to be categorical about their degree of generality – although anecdotal evidence and advertisement data confirm that pay levels are higher in the south-east region.

Appendix 6.2 : Household expenditure on waged domestic labour

	North-east	South-east
Annual household income of employers after deductions (i.e., disposable income)[1]	£30–39,000	>£40,000
Annual earnings of nanny[2]		
Take-home	£3,840	£5,760
Gross	£4,800	£7,200
Mean annual earnings of cleaners	£1,084	£940
Proportion of annual household disposable income spent on nanny[3]	13% (16%)	14% (18%)
Proportion of annual household disposable income spent on waged domestic labour	17% (20%)	16% (20%)

Sources: North-east and south-east workplace surveys, fieldwork interview data.
Notes: 1 Modal class.
2 These figures are based on the mode (£80 per week in the north-east and £120 per week in the south-east).
3 The first figure is calculated on the basis of take-home earnings, the second on gross earnings.

NOTES

1: INTRODUCTION: WAGED DOMESTIC LABOUR IN CONTEMPORARY BRITAIN

1 The question of the meaning of the term 'middle-class' and its relation to the 'service class' is considered in Chapters Two and Four. Given the concerns of this volume, we use both terms interchangeably, and in a largely descriptive fashion.

2 High-profile media interest has also characterised waged domestic labour beyond Britain, notably and recently in connection with President Clinton's search to appoint a US Attorney General, dubbed by the press, 'Nannygate'. Zoe Baird's admission that she employed undocumented waged domestic women and her non-payment of social security taxes cost her the position of Attorney General, and also led to the deportation of her Chilean employees (Mattingly, 1993).

3 The issue of the international migration of domestic labour falls outside the remit of this research, although clearly such issues are worthy of much further investigation.

4 The advertisement placed by the Qulis read: 'Mature person required to do light housework and look after 6 month baby in East Ham area. £50pw' (*Daily Telegraph*, 15 August 1992). From this it is clear that the Qulis were seeking someone to come into their home; that they wished this person to do some housework and that they were offering pay at the very bottom of the range for childcare workers. Such a job description is a very long way from that for a childminder (who cares for children in her own home, as opposed to the parental home). Moreover, given both the housework specification and the pay, neither is it an advertisement for a nanny. But, since press reports suggest that Mrs Quli was to return to her job as a senior clerk in a City bank, it is inaccurate to describe the job as seeking a mother's help (although both the wages offered and the housework specification would indicate that this was what the Qulis were looking for). In our work we have classified advertisements such as this as seeking a nanny/mother's help, i.e., someone who would have sole charge of a child (or children) in the parental home but who would also be expected to do housework. We consider the confusion contained in the press reporting on the Quli story to be indicative of the extent of the confusion which exists in Britain in the sphere of waged domestic labour relating to childcare. For a full definition of the various forms of waged domestic labour in contemporary Britain, see glossary.

5 See, for example, *Daily Telegraph*, 14 August 1992. In this report, and presumably in the trial itself, emphasis is given to repeated disagreements between Beeson and Mrs Longcroft concerning what to do with the child; to personality clashes between Beeson and Longcroft; to accusations of guilt, as well as to a statement from Beeson (under cross examination) that she did not respect Mrs Longcroft as a mother.

6 *The Hand that Rocks the Cradle* (Touchstone, 1992); *Fatal Attraction* (Paramount, 1989). And see too, G. Bedell, 'Nanny dearest', *Evening Standard*, 1992 (March), which starts off with, 'Middle-class babies should come with a Government Health Warning: Working Parents Beware: nannies are strangers; they can be weird and dangerous' and elaborates on the themes of abuse, sexual perversion, drugs, HIV and affairs!

7 A further important category of waged domestic labour in Britain through the 1980s was, of course, childminders. However, our concern in this research has been with those categories of waged domestic labour employed *within* middle-class households. Although, as a result, our study represents a far from inclusive coverage of the waged domestic labour phenomenon in contemporary Britain, we feel that the emphasis here provides an important corrective to accounts which stress spatial shifts in the site of middle-class social reproduction (Watson, 1991).

8 Interview respondents for our nanny and employer categories were generated through snowballing methods. Originally we had also intended to interview twenty-five cleaners in both study areas. However, certain of the characteristics of the cleaning labour force, specifically their tendency to work alone and to be supplementing incomes derived from benefit, meant that it proved extremely difficult to find private domestic cleaners to interview. Then, once found, we had to convince these women that we were neither Department of Social Security (DSS) 'snoops', nor DSS 'shoppers'. Finding cleaners proved a painstakingly slow process, and most of our interviews were, in fact, only achieved with the help of their employers. In addition, our interview programme with cleaners revealed remarkable consistency from the start. Both sets of circumstances meant that we heavily revised our initial fieldwork target number for this category of waged domestic labour. In our original proposal we had also intended interviewing twenty-five employers of nannies and twenty-five employers of cleaners in both areas. In practice, however, we soon found that a significant proportion of middle-class households actually employ both categories of workers. The numbers of employers interviewed in each area therefore represents the number of households needed to satisfy the target of twenty-five nanny employers and twenty-five cleaner employers. The total number of interviews we conducted was 139. All north-east interviews were conducted by Nicky Gregson; the south-east interviews were conducted by Michelle Lowe and Sarah James.

9 We do not consider it appropriate to read anything into these differences! However, it is interesting to note that in our survey work it was – in both areas – overwhelmingly the *female* partner who filled in our questionnaires!

10 As will become clear in Chapter Two, our analysis of advertisement data from *The Lady* magazine confirmed our initial research expectations. Berkshire is shown to comprise the second most important source outside London of advertisements for waged domestic labour, whilst Tyne and Wear is near the bottom of the table of advertised demand by county.

2: THE GEOGRAPHY OF WAGED DOMESTIC LABOUR IN BRITAIN IN THE 1980s

1 Later in the interview, she explains that she 'folded up' the nanny side of the agency owing to concern over the client–nanny–agency relation.
2 Both quotations are derived from interview data.
3 We discuss some of the problems with advertisement data in the following section. Although it would have been infinitely preferable to have worked from absolute employment figures on waged domestic labour, in practice both the occupational classification and Census of Employment statistics make it impossible to produce such statistics. Thus, although order 068.1 enables the isolation of the number of domestic housekeepers and cook-housekeepers working in private households, order 068.3 amalgamates the particular categories of waged domestic labour which we are most interested in (nannies and cleaners) with those working as cleaners in hotels and hospitals; playgroup supervisors; assistant teaching staff; companions, etc. In addition, although included in the Standard Industrial Classification codes, 'private domestic service' (MLH = 891) is not counted in the Census of Employment. This means that no accurate employment data on waged domestic work within private households exist, either generally or in relation to the two major categories of such labour in contemporary Britain, namely nannies and cleaners.

The importance of household advertisements was confirmed by both our regional newspaper analyses and interviews with employment agencies and home-cleaning services. Thus, whilst agencies and firms concentrating exclusively on home cleaning admitted to using the local press to find the majority of their domestic staff, advertisements emanating from such sources constituted only 22 per cent and 6 per cent respectively of advertisements recorded in our north-east and south-east newspaper surveys. Moreover, interviews with home-cleaning firms suggested that the numbers employed by them were small (typically in the range of 10 to 15) and that the number of regular clients per firm was in the range of 50 to 100.

4 Corroboration for our choice of newspapers was provided by the interview stages of our research. Many of the employers interviewed by us turned out to have used these papers for placing advertisements, whilst nannies and prospective nannies bought them in order to look for vacancies. Undeniably, though, these papers incorporate a spatial bias. For example, the *Newcastle Evening Chronicle* is a Tyneside-based paper. Although the obvious location for advertisements for domestic vacancies in the Newcastle, Gateshead and Coast areas of the conurbation, it would be a less obvious choice for households in the west and south of the region (ie., Hexham, Corbridge and the Tyne Valley, and Durham/Darlington respectively). In these cases the *Hexham Courant* and the *Northern Echo* would be more obvious papers within which to place advertisements for domestic workers. Similarly, in the south-east, the *Reading Chronicle* constitutes the obvious location for households in Reading itself, but not for those households living in either the Thameside settlements or the Wokingham area, where the *Henley Standard* and the *Wokingham Times* would be clear first choices.

5 A full record would have required us to survey all regional papers and to have amalgamated and cross-checked these for duplicate advertisements. Given the limitations of advertisement data anyway, and our consequent intention to establish broad (rather than absolute) trends, we considered this possible course to be unproductive in terms of its envisaged returns.

6 Since accurate records were not kept by *The Lady* of the number of advertisements placed per issue for domestic vacancies, it proved impossible to obtain a 10 per cent sample of all such advertisements. We therefore had no option but to use the structure of the data source to generate our sample. Since a sampling framework based on advertisement totals would have been unable to differentiate duplicate advertisements, we consider this to have been as problematic in its own way as our eventual choice of sampling framework. Indeed, pilot work on *The Lady* advertisements as a potential data source suggested that duplication of advertisements was common practice across consecutive issues.

7 The basis for this analysis has been postcode *districts* (for example, NE2, SW18, NW3). This choice was made because of the tendency of households to specify postcode districts in advertisements. Although postcode sectors are liable to change, postcode districts are relatively robust spatial units, and postcode districts within London, Newcastle and Reading are known not to have changed in the period 1981–91 (Post Office, personal communication).

8 Although advertisement duplication was common practice with *The Lady* data over two-week periods, pilot work suggested this not to be so with periods of more than one month.

9 If such advertisements specified the care of young pre-school-age children and a working week of more than fifteen hours (i.e., over five mornings/afternoons per week) they were allocated to the category 'mother's help'. Alternatively, if they contained phrases such as 'one or two mornings a week', and made no reference to childcare they were allocated to the cleaner category. These strategies accounted for almost all cases of advertisements seeking 'domestic help'.

10 Unless containing evidence to the contrary, advertisements specifying a nanny were left untouched. But those advertisements which, for example, requested a nanny, but which made it clear that the mother would be around, were changed to the category 'mother's help'. Those which specified housework in conjunction with childcare, but which suggested that the mother would not be around during the day, were reclassified as seeking a nanny/mother's help. Ambiguously worded advertisements for nanny/mothers' helps were left unaltered; the argument for this being that if the household concerned was unsure about the distinction between these two categories of waged domestic labour, then it was highly likely that the job itself would involve both childcare and housework.

11 The severity of the recession of the early 1990s, and its effect on waged domestic labour, only began to become apparent in the course of our south-east employer interviews in the summer of 1991. By then the north-east fieldwork had been completed. It was therefore impossible for us to consider fully the implications of this radically altered economic context for the employment of waged domestic labour by the middle classes. Obviously, we would wish to consider this in the future, but the indications seem to be that certain categories of waged domestic labour (particularly nannies), although influenced by recession, are proving to be resilient at least in part.

12 It needs to be recognised that maps of owner-occupation and of those working in professional/managerial occupations are based on ward data, and not on postcode districts.

13 Once more, the problem of the lack of correspondence between postcode districts and wards needs to be acknowledged. In the cases of Gosforth and Jesmond, the ward data on which we have based our comments pertain to

South Gosforth and Jesmond respectively. In the case of South Gosforth, the ward includes most of the NE3 postcode district, with the exception of the easternmost area of Kenton ward, parts of Grange ward and areas of Fawdon. Likewise, in the case of Jesmond ward, this covers most of the NE2 postcode district, with the exception of parts of Sandyford ward. In selecting the most appropriate wards to use to discuss the socio-economic characteristics of those postcode districts responsible for the majority of north-east advertised demand, we have employed a certain degree of 'triangulation'. Thus, we used both the location of our interviewed employing households, as well as the socio-economic characteristics of employing households (as gleaned from survey work), to guide us in our selection of wards.

14 This pilot work consisted of a preliminary questionnaire survey of fifty households employing waged domestic labour in both Newcastle upon Tyne and Reading. Our concerns were to establish the characteristics of employing households and the main categories of waged domestic labour employed.

15 Such findings are broadly in accordance with those from a recent Gallup survey which found that 1 million out of 19 million households in Britain employ paid domestic help regularly (*The Guardian*, 6 February 1990).

16 These findings concerning cleaners provide a marked contrast with the 'visibility' of the cleaner category amongst our advertisement data analyses, and are corroborated both by previous market research and by our interviews with employing households. The lack of 'visibility' of household advertisements for cleaners is hardly surprising given the 'undeclared' nature of this form of work (see Chapters Five and Seven).

17 Although no straightforward distinction was revealed between nanny-employing households working in the public sector and those in the private sector, those working in the public sector were more likely to make the case for either the greater provision of collective childcare facilities or greater choice in forms of provision than those employers working in the private sector. For the latter, it was the tax situation regarding childcare which came in for the greatest degree of criticism.

18 Fieldwork survey, 1990. For full details, see Chapter Five.

19 Although it might be anticipated that difficulties in labour availability in the south-east might 'feed through' into employer–employee practices, we found no evidence to support such an argument.

3 PERSPECTIVES ON WAGED DOMESTIC LABOUR

1 *Upstairs, Downstairs* (London Weekend Television, 1973); *You Rang, M'Lord?* (British Broadcasting Corporation, 1991); *Jeeves and Wooster* (Granada, 1990); C. Brontë, *Jane Eyre*, London, Collins, 1969; I. Compton-Burnett, *Manservant and Maidservant*, London, David & Charles, 1947; M. Forster, *Lady's Maid*, Harmondsworth, Penguin, 1947; C. Cookson, *The Mallen Girl*, London, Heinemann, 1974; and the following, also by Cookson – *Hamilton*, London, Heinemann, 1983; *The Black Velvet Gown*, William Heinemann, London, 1984; *The Black Candle*, London, Bantam, 1989; *Kate Hannigan*, London, Futura, 1989; *The Gillyvors*, London, Bantam, 1990.

2 One of the pockets of servant concentration within the provincial cities was Gosforth (349 servants to 1,000 families), a point which is further evidence of the long-standing importance of this area of Newcastle upon Tyne as an area of middle-class housing (Robinson, 1988).

3 The other major area of historical investigation is France (see, for example, Fairchilds, 1979, 1984; McBride, 1974, 1976; Maza, 1983).

4 UNPACKING DEMAND

1 See, Vax Home Cleaning Report (1991). The survey conducted by MORI was a representative quota sample of 1,062 women aged 18-plus, across sixty-eight constituency sampling points in late 1990. The data are not disaggregated by social class.

2 There is, moreover, evidence to suggest that cleaning tasks themselves are becoming increasingly fragmented as manufacturers expand their product ranges.

3 'Proper' household cleaning refers to thorough, as opposed to superficial, cleaning, i.e., to activities such as moving furniture when vacuuming.

4 These findings are also confirmed by various market research surveys. For example, a 1990 Gallup survey suggested that 62 per cent of all women do all the household dusting, cleaning and vacuuming; that 1 per cent of all men do all the household washing; and that 2 per cent of all men do all the household ironing compared to 87 per cent of all women. Similar findings are reported from the MORI survey (Vax Home Cleaning Report, 1991).

5 Although the MORI survey is not disaggregated by class, it provides some significant findings concerning the differences between women in full-time employment, those in part-time employment and those who are full-time housewives. 21 per cent of those women surveyed who were in full-time employment claimed to spend over ten hours per week cleaning (as against 43 per cent of those employed on a part-time basis and 45 per cent of those who were full-time housewives).

6 See *The Guardian* (23 September 1992) for a preliminary report from the 1991 General Household Survey.

7 Although it could be argued that it is possible for working-class households to employ paid domestic help, in practice the material resources available to such households would preclude this. Some indication of the material resources available to our respondent households, and the amount of disposable income devoted to expenditure on waged domestic labour, is given in Appendix 6.2.

8 It was, however, suggested to us during our interviews with employers that kin relatives were occasionally available to such households on an irregular basis, typically around the time of major events such as childbirth.

9 All names and household names used henceforward are pseudonyms. In addition, various biographical and locational details have either been altered or omitted to preserve respondent confidentiality.

10 For couples with two or more pre-school-age children, the cost advantages of employing a nanny relative to a childminder are considerable. This is because parents are charged on a place basis by childminders. Childminder rates are at the childminder's discretion and can range from £1.30 to £2.25 per hour. The National Childminding Association (NCA) recommends a minimum of £50 per week full-time per child. For households with more than one pre-school-age child it is therefore cheaper to employ a nanny. In comparison, whilst most of the nannies whom we interviewed had been given pay rises with the arrival of another child, such rises were a small fraction of their previous weekly earnings.

11 Although, as in the case of the Davieses, it was the relative inflexibility of

childminders' hours which our respondents pointed to, a minority also commented on the quality of care offered by childminders. In this respect, the Lyleses are not untypical (and see too the Joneses and the Keiths). James Lyles is an architect and Stephanie Lyles a director of a large pharmaceutical company. They have two children, one of whom is pre-school age. Here they both outline their feelings about childminders:

JL: We looked at childminders and at the time we were very unimpressed with what we saw. We went to two or three baby childminders and it was just awful. Had there been a terrific childminder – and I'm sure that there must be lots about that we didn't see – but the ones we saw were unimpressive . . .

SL: The one I remember vividly – and she sounded good because she had fostered children – and I thought she must be good with kids, but she'd actually fostered difficult teenagers and she was a safe house for runaways, and she had a pet rottweiler. She said she sometimes got angry parents turning up on the doorstep and we thought with angry parents and a rottweiler wandering around it wasn't what we wanted.

12 When we explored the relative invisibility of waged domestic labour amongst the middle-classes our respondents typically referred to one or more of the following: guilt; feelings of inadequacy (ie., that one should be able to cope either single-handedly or as a partnership with the demands of domestic labour and full-time service class employment); and the association of waged domestic labour with the upper classes. Such references suggest that whilst those elements of the new middle-classes which employ waged domestic labour have rationalised at a personal level the use of such forms of working, they are far from convinced of its general social acceptability in contemporary Britain.

5 UNPACKING SUPPLY

1 See for example, T. Kelsey, 'Student's dream wrecked by domestic slavery', *The Independent* (4 January 1992). Kelsey estimates that there may be as many as 20,000 migrant women of colour working as waged domestics in private households in contemporary Britain, but no Home Office statistics are available. These workers enter Britain as the domestic staff of specified immigrants, are granted visitor's visas and have no legal status independent from their employers. Should they cease working for this individual and/or household, they are liable to be deported.

2 Although it is, of course, impossible for us to say much from our research about this situation, it appears that the differences between the two areas in terms of labour supply reflect both contrasting local labour market conditions through the 1980s and the different role of the north-east and the south-east within the spatial division of labour. Thus, in the case of the north-east we would anticipate that two factors proved particularly important in ensuring the ready availability of waged domestic labour, namely, the type of 'female' employment generally available within the region and high levels of male unemployment through the 1980s. In contrast, in the booming south of the 1980s, with jobs relatively easy to come by (particularly within the service sector) and low levels of male unemployment, the alternatives to waged domestic work would almost certainly have appeared more attractive.

3 In contrast to our employer households, we use first-name pseudonyms for our waged domestic workers. This is to help readers differentiate between employing households and waged domestics. We wish to stress that no other interpretation should be placed on this device.

4 That is, a product of the private Norland Nursery Training College, Berkshire, for nannies.

5 Although, in theory the NNEB course is 'open entry', in practice many of the technical colleges are heavily over-subscribed. Courses validated for fifty places per annum, for example, would regularly expect to get between 250 and 375 applications for their places (interview data). Given this, course tutors frequently set entry 'tests' (to establish basic literacy levels) and then interview.

6 The sixty-eight returns were made up of thirty-four from each of the surveyed north-east and south-east colleges. Originally we had hoped to survey all course participants in both colleges (i.e., approximately 100 in the two years in the north-east college and 108 in the south-east). However, our access to course students was controlled by course tutors. We could therefore only distribute questionnaires to students during one particular class. The completed returns represented a 100 per cent return from those attending these particular classes.

7 See, for example, J. Coles, 'Job "juggling act" takes toll of mother and child', *The Guardian* (1 March 1990).

8 R. Chantry Price (Director: National Nursery Examination Board), personal communication.

9 Evidence suggests that the national picture contains considerable geographical variation. As part of our research programme we attempted to get statistical returns from all the courses validated for NNEB students in our two study areas. Unfortunately, the response rate from course tutors was extremely variable. Indeed, in the north-east study area, the colleges responding either did not correspond with those areas known to produce the highest levels of demand for nannies or had only just been validated. However, the information we have from the Reading area over the two years 1988–9, shows that 67 per cent (fifty-seven students) of those leaving Reading Technical College with the NNEB qualification found employment as nannies. In comparison, in the equivalent period 33 per cent (twenty-three students) of those leaving Monkwearmouth College (Sunderland) went into this form of paid work. Since Sunderland is not an obvious source of nannying jobs in the north-east, we would not wish to infer anything from such results! However, that 67 per cent of those leaving Reading Technical College with the NNEB found this form of employment is indicative of the levels of demand for this form of waged domestic labour in the Reading area in the late 1980s.

10 In both the north-east and south-east, approximately 68 per cent of students surveyed responded that they were considering employment as a nanny. Roughly comparable percentages were considering working in a school (44 per cent in the north-east, 34 per cent in the south-east). Under 10 per cent in both areas were thinking about either hospital work or work in local authority nurseries. Whilst some students clearly were considering more than one option, it is significant that nannying was being considered by almost twice as many students as the next most popular option.

11 See, for example, C. Jardine, 'Nanny moves into the office', *Sunday Telegraph* (29 March 1990). Jardine argues that the predicted growth in workplace

nurseries in the 1990s will lead to an expansion in employment openings in private sector day care for NNEB holders, although she notes the shortfall in the number of nursery nurses trained to manage such nurseries.

6 NANNY EMPLOYMENT IN CONTEMPORARY BRITAIN

1 Although we did not investigate the London nanny-scene directly, various aspects of our investigations confirmed the predominantly live-in nature of nanny employment there. These included interviews with nannies and with various employment agencies, as well as the advertisement analysis reported in Chapter Two.

2 'Bounce' and 'Tumble Tots' are gymnastic activities for young children: Nanny/mother and toddler groups are voluntary sector informal social meetings which anyone with a toddler can attend for a small fee (payable on entry). Playgroups (for children of 2 to 3 plus) are also predominantly voluntary sector activities, but have a fixed number of reserved places and entail leaving the child. They are normally used as a precursor to nursery school. As important as these non-home-based activities are visits to (and by) other nannies and their charges. The normal social practice by far within nanny employment is the operation of the nanny circle. Comprising a minimum of two to three nannies (and their charges) and a maximum (at least in our study areas) of fifteen, the nanny circle fulfils a number of roles. For nannies, such groups provide a means to 'get away' from children's talk, a means to give vent to grievances, and the basis for the exchange of information pertaining to the labour market. For the 'charges', such circles constitute the basis for wider socialisation and friendships. As with their equivalents – informal mothers' networks – the basis of these networks lies in reciprocity and helping out. Thus, nannies usually take it in turns to play host to the others in the circle and frequently help one another out, for example by minding another nanny's charges whilst she pops out.

3 The attitudes to motherhood of the female partners we interviewed is a vast area to which it is impossible to do justice here. Suffice it to say that, whilst almost all of those we interviewed saw full-time motherhood as an unattractive proposition, they also articulated many of the ideas, conflicts and contradictions reported by Brannen and Moss (1991). Of particular importance for many of the women were the predictable feelings of guilt. However, many also stressed unsupportive/unsympathetic and in some cases hostile work colleagues/contexts.

4 The following comments, from one course tutor, are typical of the way in which the NNEB course is seen to offer professional training in childcare:

It's a professional qualification. And we say to the girls, we very much say from the beginning, that they are professionals. . . . I interviewed a girl once, and her mother was with her, and she said, 'Well I brought seven up. Why does she need to go to college for two years to learn that?' That's all right. But it's, 'Why are they doing this?' 'Why do they play in this way?' 'Why should I provide this for them at this time?'. 'What is going wrong at this stage?' So they recognise the normal. We teach 'the normal' in great depth, so they will recognise the abnormal.

5 During our fieldwork we encountered little evidence that the NNEB qualified nanny commanded a higher wage than the non-qualified nanny in the labour market. Indeed the highest paid nannies in both of our study areas were, in

fact, *unqualified* girls employed in nanny-shares. We therefore are very firmly of the view that nanny wages within any local area are very much dictated by 'a going rate' (i.e., by what parents are willing, and can afford, to pay). As a consequence, nannies simply have the option of deciding whether (or not) they are willing to work for a certain amount. Given the unregulated nature of the nanny form of waged domestic labour, this degree of employer control over wages is unsurprising. We also encountered clear signs that employers operate (or at least try to operate) a wage standardisation policy within local areas. Thus, most of our employers referred to 'asking around' as to the *standard rate* before employing a nanny, and a number made it clear that they took a dim view of employers who paid above this going rate. Our discussions also left us in no doubt that the most interesting information which *we* could provide employers concerned comparative wage levels! In all this, of course, *standardisation and not differentiation according to qualification/experience* are being emphasised.

6 Another influence, which we did not investigate directly, but which would seem to have considerable potential for shaping day-to-day arrangements as between women is the variety of 'working mother' guides. In many senses the late twentieth-century middle-class woman's equivalent of the nineteenth-century household manuals, these volumes offer practical advice and hints on the problem areas of working motherhood. Here, for example, are some characteristic comments on the nanny from Litvinoff and Velmans (1987: 225–7), which resonate strongly with many of the issues discussed here and in Chapter Four:

> For those who can afford it, employing a nanny ... is an attractive option. It means leaving your children in the security of their own home, surrounded by their own toys, favourite foods, friends, in the care of someone whose function is to give them her full, personal attention. *She is accountable only to you, so that you can expect her to follow your instructions about the way your children are looked after, disciplined, fed and entertained* [our emphasis]. It also has the advantage of convenience to you: it means you don't have to fit rushing to drop off your children and pick them up again into a hectic working day. There will also be less pressure on you to get home exactly on time, and there is the possibility of built-in babysitting or housework as well! ... She is not a servant but an employee and she is respected as a professional, having either been trained as a nursery nurse or had a number of years experience of looking after small children ... If you feel guilty at the thought of employing a nanny because it doesn't fit in with your social principles tell yourself you are providing a much needed job and helping someone else off the dole.

The same section also includes an illuminating characterisation of the 'ideal nanny'; advice on how to find her; recruitment and how to be an employer.

7 Enforced housework was the second most commonly cited grievance articulated by the nannies we interviewed, after pay. The reason should be obvious; the entire nanny occupation rests on the separation of childcare from general household chores. However the conjoining of these tasks in the form of unwaged domestic labour exerts a major influence on the actions of many nannies. Importantly, though, housework (and its performance in such circumstances) definitely takes the form of a favour, that is, something which is never to be expected. Some idea of the resentment which nannies can feel over compulsory housework is conveyed in the following extracts

It was 'Cheryl do this' and 'Cheryl do that'; Cheryl wash this' and 'Cheryl wash that'. I left after 3 days.

It became Amanda, would you mind vacuuming out the bathroom after you've given Jamie his bath?' 'Amanda, would you do the kitchen floor – it's a bit dirty this week?' And then the ironing! This 'occasional shirt and blouse' turned out to be the whole family's ironing! . . . so one day I said to her 'Look, this isn't in my contract. I didn't mind doing it at first when Jamie was younger and when he slept a lot. But now he doesn't sleep as much and I haven't got as much time to do the ironing.' And she said, 'In your contract it says family ironing.' And I said, 'But we came to a verbal agreement'. But I had no come-back. She said, 'If it's time to do the ironing just put him in his play pen.' And I was thinking 'Mrs Jones' [NNEB tutor] would turn in her grave.' That's when I went off and started looking for another job.

7 CLEANER EMPLOYMENT IN CONTEMPORARY BRITAIN

1 Our interview programme with cleaners suggested twenty hours to be the absolute ceiling on labour force participation in this form of waged domestic labour. There are several possible reasons why, including the physical work burden of combining multiple-employer cleaning with unwaged domestic work; the greater risk of being 'shopped' to the DSS; the loss of flexibility; and the fact that beyond twenty hours takes one into the realms of other possible forms of part-time labour force participation.

8 THEORETICAL AND POLITICAL REFLECTIONS

1 Such forms of cross-class subsidy do not, of course, apply in circumstances where nannies are employed in a live-in capacity.
2 We explore such arguments in more depth elsewhere (see Gregson and Lowe, forthcoming a).
3 One of the reasons why we found ourselves sucked in by these arguments is an interesting reversal of the accepted power relations of social research. Thus, rather than feeling able to express our interpretations of the politics of waged domestic labour in contemporary Britain, we – the researchers – both felt imprisoned by the overt and covert criticisms of our representations by many of our middle-class *female* respondents. We explore this situation at length elsewhere, see Gregson and Lowe, forthcoming b).

GLOSSARY

Au Pair Officially, according to Home Office regulations until recently, an au pair is a single girl, without dependants. She is usually between 17 and 27, from a western European country, and comes to live with a family to learn English. In return for board, lodging, some pocket money and the opportunity to attend English classes, she can be expected to do a maximum of five hours light domestic work per day, for example, vacuuming, ironing, dusting, washing up, loading and unloading the washing machine, some shopping, perhaps some cooking, and childcare; plus two or three hours babysitting a week. An au pair would expect to have one full day a week free. An au pair can stay in Britain for a maximum of two years.

Butler A butler's traditional tasks have been to organise and supervise the other domestic staff employed within a household; to control the ordering, storing and serving of wines and spirits; and to attend to the personal needs, personal effects and wardrobe of his (male) employer(s). The position is usually residential. At the apex of the nineteenth-century hierarchy of domestic service occupations, the contemporary butler may still be expected to perform the tasks listed above. However, butlers may also be expected 'to run homes, organise entertaining, plan menus and shop, amuse the children, make travel arrangements – even fly the family jet' (Higgins, 1991). Many are the product of the Ivor Spencer International School for Butlers/ Personal Assistants, London.

Carer A carer's chief tasks are to provide companionship and perform personal services for the employer(s). Increasingly, carers are employed to care for elderly or disabled people in their own homes. They may live in the employer's home. Alternatively, they may live out and come to work on a daily basis.

Cleaner A cleaner's chief task is to clean the interior of private households. However, cleaners may also do the washing and/or ironing as well. Cleaners usually work either one or two sessions per week for one household, and may also work for more than one employer. One session is usually three to four hours in duration. Cleaners may work independently. Alternatively, they may be employees of firms specialising in household cleaning services.

Cook A cook plans menus and prepares and cooks food in private households. A cook usually holds a residential position but may also be hired on a weekly or even a daily basis.

Couple This is usually a husband/wife team who work and live together, either on the employer's land (for example, in a 'tied' house) or on the

305

employer's premises (for example, in a self-contained flat). The accommodation is part of the job. A clear division of labour exists within this form of waged domestic work. The husband, therefore is usually employed as a gardener/handyman/driver, whilst the wife works as a cook/housekeeper.

Gardener A gardener cultivates flowers, trees, shrubs and other plants and maintains lawns, paths and patios in private gardens. Gardeners are usually employed on a casual/part-time basis. Payment is by the hour, cash in hand.

Housekeeper A general domestic worker whose tasks may include anything pertaining to the day-to-day reproduction of the household (ie., cleaning, cooking, washing, ironing, shopping). The housekeeper is not normally required to do childcare related tasks/activities in relation to young (pre-school-age) children, but may be required to perform this role with older children. The position may be either live-in or live-out, full-time/part-time.

Maid Cleans rooms, prepares food and serves meals, washes dishes and performs additional domestic duties in private households. The position can be residential or daily.

Maternity Nurse A maternity nurse is someone who specialises in the care of very young babies. She will either be a fully trained nurse, or a nursery nurse (NNEB) with a lot of experience of infants. A maternity nurse is usually employed on a short contract – usually from four to six weeks, but for a maximum of about three months. However, maternity nurses can be hired by the week – sometimes even by the day. A maternity nurse will expect to do night duty with small infants – but will be paid a lot more and will obviously live in. They usually earn twice as much as the average nanny. The maternity nurse is on duty twenty-four hours a day, six days a week, the role being to establish a good routine for the baby.

Mother's Help The mother's help is employed to assist the mother in general housework and childcare. In such instances the employer *is not* in waged employment. Unlike the nanny, the mother's help is not usually trained and/or qualified. Such positions may be either live-in or live-out, full-time/part-time.

Nanny A nanny's tasks are to look after the (young) children within a household in the absence of the employer who is in either full-time or part-time employment). The tasks include basic childcare activities (bathing, feeding, clothing etc.) and entertaining/educating children. A nanny may be either live-in or live-out. Nannies are mostly the product of technical colleges (the standard qualification is the National Nursery Examination Board (NNEB) certificate). However, they may also be trained through one of the private nanny training colleges (ie., Norland, Chiltern, Princess Christian).

Nanny/mother's help As above, but the nanny/mother's help is also expected to do general housework.

Nanny Share A nanny-share is a nanny employed by two, or more, families (usually no more than three) and involves the combined care of these households' (young) children. The nanny may work in either house or may alternate between the houses of employers. A nanny working for more than one household would expect to earn considerably more than one working for one household only.

BIBLIOGRAPHY

Abel, R.L. (1989) 'Between market and state: the legal profession in turmoil', *The Modern Law Review* 52, 3: 285–325.

Abercrombie, N. and Urry, J. (1983) *Capital, Labour and the Middle Classes*, London: George Allen & Unwin.

Allen, J. (1988a) 'Service industries: uneven development and uneven knowledge', *Area* 20, 1: 15–22.

—— (1988b) 'Towards a post-industrial economy?' in J. Allen and D.B. Massey (eds) *The Economy in Question*, London: Sage.

—— (1988c) 'The geographies of services', in D.B. Massey and J. Allen (eds) *Uneven Redevelopment*, London: Hodder & Stoughton.

Anderson, C.A. and Bowman, M.J. (1953) 'The vanishing servant and the contemporary status system of the American South', *American Journal of Sociology* 59, 2: 215–30.

Aubert, V. (1956) 'The housemaid – an occupational role in crisis', *Acta Sociologica* 1, 3: 149–58.

Bagguley, P. and Mann, K. (1992) 'Idle thieving bastards: scholarly representations of the underclass', *Work, Employment and Society* 6, 1: 113–26.

Bagguley, P., Mark Lawson, J., Shapino, D., Urry, J., Walby, S. and Warde, A. (1990) *Restructuring: Place, Class and Gender*, London: Sage.

Bailyn, L. (1970) 'Career and family orientation of husbands and wives in relation to marital happiness', *Human Relations* 22: 97–113.

—— (1978) 'Accommodation of work to family', in R. Rapoport and R.N. Rapoport (eds) *Working Couples*, New York: Harper & Row.

Balderson, E. (1982) *Backstairs Life in an English Country House*, Newton Abbot: David & Charles.

Banks, J.A. (1954) *Prosperity and Parenthood: A Study of Family Planning Among the Victorian Middle Class*, London: Routledge & Kegan Paul.

Barlow, J. and Savage, M. (1986) 'The politics of growth: cleavage and conflict in a Tory heartland', *Capital and Class* 31: 157–82.

Bebbington, A.C. (1973) 'The function of stress in the establishment of the dual-career family', *Journal of Marriage and the Family* 35, 3 (August): 530–7.

Beechey, V. and Perkins, T. (1987) *A Matter of Hours*, Cambridge: Polity.

Bell, C. (1968) *Middle Class Families*, London: Routledge & Kegan Paul.

Bell, D. (1973) *The Coming of Post-Industrial Society*, London: Heinemann.

Birch, B. (1984) 'The sphinx in the household? A new look at the history of household workers', *Review of Radical Political Economy* 16, 1: 105–20.

Black, C. (ed.) (1915) *Married Women's Work: Report of an Enquiry Undertaken by the Women's Industrial Council*, London: G. Bell.

Bone, M. (1978) *Pre-school Children and the Need for Day Care*, London: HMSO.

Bonney, N. (1988a) 'Dual earning couples: trends of change in Great Britain', *Work, Employment and Society* 2, 1: 89–103.

—— (1988b) 'Gender, household and social class', *British Journal of Sociology* 31, 2: 28–46.

Booth, C. (1903) *Life and Labour of the People in London*, London: Macmillan.

Boserup, E. (1970) *Women's Role in Economic Development*, London: Allen & Unwin.

Bourdieu, P. (1990) *In Other Words*, Cambridge: Polity.

Bowlby, J. (1951) 'Maternal care and mental health', *Bulletin of the World Health Organization* 3: 355–534.

—— (1958) 'The nature of the child's tie to his mother', *Journal of Psychoanalysis* 39: 350–73.

—— (1965) *Childcare and the Growth of Love*, Harmondsworth: Penguin (second edition).

Branca, P. (1975) *Silent Sisterhood: Middle Class Women in the Victorian Home*, London: Croom Helm.

—— (1976) 'Image and reality: the myth of the idle Victorian woman', in M. Hartmann and L. Banner (eds) *Clio's Consciousness Raised*, New York: Octagon Books.

Brannen, J. and Moss, P. (1988) *New Mothers at Work: Employment and Childcare*, London: Unwin Hyman.

—— and —— (1991) *Managing Mothers: Dual Earner Households after Maternity Leave*, London: Unwin Hyman.

—— and Wilson, G. (1989) *Give and Take in Families: Studies in Resource Distribution*, London: Unwin Hyman.

Brenner, J. and Laslett, B. (1986) 'Social reproduction and the family', in U. Himmelstrand (ed.) *Sociology; From Crisis to Science, Volume 2: The Social Reproduction of Organisation and Culture*, London: Sage.

Bresse, C. and Gomer, H. (1988) *The Good Nanny Guide*, London: Century Hutchinson.

Britten, N. and Heath, A. (1983) 'Women, men and social class', in E. Gamarnikow *et al.* (eds) *Gender, Class and Work*, London: Heinemann.

Brookfield, H. (1975) *Interdependent Development*, London: Methuen.

Broom, L. and Smith, J.H. (1963) 'Bridging occupations', *British Journal of Sociology* 14: 321–34.

Bryan, B., Dadzie, S. and Scafe, S. (1985) *The Heart of the Race*, London: Virago.

Buck, N. (1985) 'Service industries and local labour markets: towards an anatomy of service employment', Regional Science Association Annual Conference, September.

Bunster, X. and Chaney, E.M. (1985) *Sellers and Servants: Working Women in Lima, Peru*, New York: Praeger.

Burnett, J. (1974) *Useful Toil: Autobiographies of Working People from the 1820s–1920s*, London: Allen Lane.

Butler, C.V. (1916) *Domestic Service*, London: G. Bell.

Callahan, H.C. (1977/8) 'Upstairs-downstairs in Chicago, 1870–1907', *Chicago History* 6, 1: 195–209.

Carter, M.J. and Carter, S.B. (1981) 'Women's recent progress in the professions or, women get a ticket to ride after the gravy train has left the station', *Feminist Studies* 7, 3: 477–504.

Cavendish, R. (1982) *Women on the Line*, London: Routledge & Kegan Paul.

Chaney, E. (1985) 'Agripina', in X. Bunster and E.M. Chaney *Sellers and Servants*, New York: Praeger.

—— and Garcia Castro, M. (eds) (1989) *Household Workers in Latin America and the Caribbean*, Philadelphia: Temple University Press.

Chaplin, D. (1964) 'Domestic service and the Negro', in A.B. Shostak and W. Gomberg (eds) *Blue Collar World: Studies of the American Worker*, Englewood Cliffs: Prentice Hall.

—— (1978) 'Domestic service and industrialization', in R. Thomasson (ed.) *Comparative Studies in Sociology*, Greenwich, Conn.: Jai Press.

City of Newcastle upon Tyne Policy Services Department (1983) *Newcastle upon Tyne 'City Profiles'. Results from the 1981 Census*.

Clark-Lewis, E. (1985) '"This work had an end": the transition from live-in to day work', in *Southern Women: the Intersection of Race, Class and Gender*, Center for Research on Women Working Paper No. 2, Memphis: Memphis State University.

Clarke, D.G. (1974) *Domestic Workers in Rhodesia: the Economics of Masters and Servants*, Givelo: Mambo Press.

Cock, J. (1980a) *Maids and Madams: A Study in the Politics of Exploitation*, Johannesburg: Ravan Press.

—— (1980b) 'Domestics servants in the political economy of South Africa', *Africa Perspective* 15: 42–53.

—— (1981) 'Disposable nannies: domestic servants in the political economy', *Review of African Political Economy* 21: 63–83.

Cockburn, C. (1983) *Brothers: Male Dominance and Technological Change*, London: Pluto.

—— (1985) *Machinery of Dominance: Women, Men and Technical Know-how*, London: Pluto.

—— (1987) *Two Track Training: Sex Inequalities and the YTS*, London: Macmillan.

Cohen, B. (1988) *Caring for Children: Services and Policies for Childcare and Equal Opportunities in the United Kingdom*, Report for EC Childcare Network, London: Commission of the European Communities.

Cole, S. (1991) 'Changes for Mrs Thornton's Arthur: patterns of domestic service in Washington DC, 1800–1835', *Social Science History* 15, 3: 367–80.

Colen, S. (1986) 'With respect and feelings: voices of West Indian childcare and domestic workers in New York', in J. Cole (ed.) *All American Women: Lines that Divide, Ties that Bind*, New York: Free Press.

Collins, M. (1985) 'Silenced: black women as domestic workers', *Trouble and Strife* 6: 15–16.

Congdon, P. (1987) *A Map Profile of Change in London Wards*, London Research Centre Statistical Series, no. 61.

Connell, R.W. (1987) *Gender and Power: Society, the Person and Sexual Politics*, Cambridge: Polity.

Cooke, P. (1985) 'Radical regions? Space, time and gender relations in Emilia, Provence and South Wales', in G. Rees *et al.* (eds) *Political Action and Social Identity: Class, Locality and Ideology*, Basingstoke: Macmillan.

—— (ed.) (1989) *Localities*, London: Unwin Hyman.

Coser, L.A. (1973) 'Servants: the obsolescence of an occupational role', *Social Forces* 52, 1: 31–40.

Cowan, R.S. (1976) 'The industrial revolution in the home: household technology and social change in the twentieth century', *Technology and Culture* 17, 1:1–23.

—— (1983) *More Work for Mother: The Ironies of Household Technology from the Open Hearth to the Microwave*, New York: Basic Books.

Coyle, A. (1982) 'Sex and skill in the organisation of the clothing industry', in J. West (ed.) *Women, Work and the Labour Market*, London: Routledge.

BIBLIOGRAPHY

—— (1985) 'Going private: the implications of privatisation for women's work', *Feminist Review* 21, Winter: 5–24.

Crang, P. and Martin, R. (1991) 'Mrs Thatcher's vision of the "new Britain" and the other sides of the "Cambridge phenomenon"', *Environment and Planning D: Society and Space* 9: 91–116.

Croft, S. (1986) 'Women, caring and the recasting of need – a feminist reappraisal', *Critical Social Policy* 16, 6: 23–39.

Crompton, R. (1986) 'Women and the "service class"', in R. Crompton and M. Mann (eds) *Gender and Stratification*, Cambridge: Polity.

—— (1987) 'Gender, status and professionalism', *Sociology* 21, 3: 413–28.

—— (1989a) 'Class, theory and gender', *British Journal of Sociology* 40, 4: 565–87.

—— (1989b) 'Women in banking', *Work, Employment and Society* 3, 2: 141–56.

—— and Jones, G. (1984) *A White Collar Proletariat?* London: Macmillan.

—— and LeFeubvre, N. (1992) 'Gender and bureaucracy: women in finance in Britain and France', in M. Savage and A. Witz (eds) *Gender and Bureaucracy*, Oxford: Basil Blackwell.

—— and Mann, M. (eds) (1986) *Gender and Stratification*, Cambridge: Polity.

—— and Sanderson, K. (1986) 'Credentials and careers: some implications of the increase in professional qualifications amongst women', *Sociology* 20, 1: 25–42.

—— and —— (1987) 'Where did all the bright girls go?', *Quarterly Journal of Social Affairs*, April: 135–47.

—— and —— (1990) *Gendered Jobs and Social Change*, London: Unwin Hyman.

Cullwick, H. (1984) *The Diaries of Hannah Cullwick, Victorian Maidservant*, London: Virago.

Dahrendorf, R. (1959) *Class and Class Conflict in an Industrial Society*, London: Routledge & Kegan Paul.

Dalley, G. (1983) 'Ideologies of care: a feminist contribution to the debate', *Critical Social Policy* 8, 3: 72–81.

—— (1988) *Ideologies of Caring – Rethinking Community and Collectivism*, London: Macmillan.

Damesick, P.J. (1986) 'Service industries, employment and regional development', *Transactions of the Institute of British Geographers* 11, 2: 212–26.

Daniels, P. (1986) 'Producer services and the post-industrial space-economy', in R. Martin and B. Rowthorn (eds) *The Geography of Deindustrialisation*, London: Macmillan.

—— Leyshon, A. and Thrift, N. (1986) *UK Producer Services: The International Dimension*, Working Papers on Producer Services, no. 1, University of Liverpool.

Davidoff, L. (1973a) 'Above and below stairs', *New Society* 24, 551: 181–3.

—— (1973b) 'Domestic service and the working-class life-cycle', *Bulletin of the Society for the Study of Labour History*, 26: 10–12.

—— (1974) 'Mastered for life: servant and wife in Victorian and Edwardian England', *Journal of Social History* 7, 4: 406–59.

—— (1979) 'The separation of home and work? Landladies and lodgers in nineteenth and twentieth century England', in S. Burman (ed.) *Fit Work for Women*, London: Croom Helm.

—— (1983) 'Class and gender in Victorian England' in J.L. Newton, M.P. Ryan and J.R. Walkowitz (eds) *Sex and Class in Women's History*, London: Routledge & Kegan Paul.

—— and Hall, C. (1987) *Family Fortunes: Men and Women of the English Middle Classes, 1780–1850*, London: Hutchinson.

—— and Hawthorn, R. (1976) *A Day in the Life of a Victorian Domestic Servant*, London: Allen & Unwin.

Dawes, F. (1974) *Not in Front of the Servants: A Time Portrait of English Upstairs/ Downstairs Life*, New York: Taplinger.

Deem, R. (1978) *Women and Schooling*, London: Routledge & Kegan Paul.

—— (ed.) (1980) *Schooling for Women's Work*, London: Routledge & Kegan Paul.

—— (1986) *All Work and No Play: The Sociology of Women and Leisure*, Milton Keynes: Open University Press.

Dex, S. (1987) *Women's Occupational Mobility*, Basingstoke: Macmillan.

—— (1988) *Women's Attitudes Towards Work*, Basingstoke: Macmillan.

Downing, H. (1980) 'Word processors and the oppression of women', in T. Forester (ed.) *The Microelectronics Revolution: The Complete Guide to the New Technology and its Impact on Society*, Oxford: Basil Blackwell.

Dubofsky, M. (1980) 'Neither upstairs, nor downstairs: domestic service in middle-class American homes', *Reviews in American History* 8, 1: 86–91.

Dudden, F. (1983) *Serving Women: Household Service in Nineteenth-Century America*, Middletown, Conn.: Wesleyan University Press.

Duncan, R.P. and Perucci, C.L. (1976) 'Dual-occupation families and migration', *American Sociological Review*, 41: 252–61.

Duncan, S.S. (1991) 'The geography of gender divisions of labour in Britain', *Transactions of the Institute of British Geographers* 16: 420–39.

Duncan, S.S. and Savage, M. (1989) 'Space, scale and locality', *Antipode* 31: 179–206.

—— and —— (1991) 'New perspectives on the locality debate', *Environment and Planning A* 23: 155–64.

Eaton, I. (1899) 'Report on domestic service' in W. Du Bois (ed.) *The Philadelphia Negro*, reprinted New York: Schocken Books, 1967.

Ebery, M. and Preston, B. (1976) *Domestic Service in Late Victorian and Edwardian England, 1871–1914*, Reading Geographical Papers, Reading University, Department of Geography.

Edgell, S. (1980) *Middle Class Couples*, London: George Allen & Unwin.

Edholm, F., Harris, O. and Young, K. (1977) 'Conceptualising women', *Critique of Anthropology* 9/10: 101–30.

Enloe, C. (1989) *Bananas, Beaches and Bases*, London: Pandora.

Epstein, C. (1971) 'Law partners and marital partners: strains and solutions in the dual career family enterprise', *Human Relations* 24, 6: 549–64.

Fairchilds, C. (1979) 'Masters and servants in eighteenth century Toulouse', *Journal of Social History* 12, 3: 368–93.

—— (1984) *Domestic Enemies: Servants and their Masters in Old Regime France*, Baltimore: Johns Hopkins Press.

Faludi, S. (1992) *Backlash: The Undeclared War against Women*, London: Chatto & Windus.

Femiola, C. (1992) *Day Care in the Home: A Study of the Issues Relating to the Quality of Care Provided by Nannies*, London: Working Mothers Association.

Field, F. (1989) *Losing Out: the Emergence of Britain's Underclass*, Oxford: Basil Blackwell.

Fielding, A.J. (1989) 'Inter-regional migration and social change: a study of South east England based upon the Longitudinal Study', *Transactions of the Institute of British Geographers* 14, 1: 24–36.

Finch, J. (1984) 'Community care: developing non-sexist alternatives', *Critical Social Policy* 9, 3: 6–18.

—— (1985) 'The deceit of self-help: pre-school play groups and working-class mothers', *Journal of Social Policy* 13, 1: 1–20.

—— (1986) 'Community care and the invisible welfare state', *Radical Community Medicine*, Summer: 15–22.

—— (1987) 'Whose responsibility? Women and the future of family care', in I. Allen *et al.*, *Informal Care Tomorrow*, London: Policy Studies Institute.

—— (1989) *Family Obligations and Social Change*, Cambridge: Polity.

—— and Groves, D. (1980) 'Community care and the family: a case for equal opportunities?' *Journal of Social Policy* 9, 4: 487–514.

—— and Groves, D. (eds) (1983) *A Labour of Love: Women, Work and Caring*, London: Routledge & Kegan Paul.

—— and Land, H. (1981) *Parity Begins at Home*, London: EOC/SSRC.

—— and Mason, J. (1990) 'Gender, employment and responsibilities to kin', *Work Employment and Society* 4, 3: 349–67.

Finn, D. (1987) *Training Without Jobs: New Deals and Broken Promises*, Basingstoke: Macmillan.

Fogarty, M.P., Rapoport, R. and Rapoport, R.N. (1971) *Sex, Career and Family*, London: Allen & Unwin.

Fox, B. (ed.) (1980) *Hidden in the Household: Women's Domestic Labour under Capitalism*, Toronto: Women's Press.

Frank, A.G. (1967) *Capitalism and Underdevelopment in Latin America*, New York: Monthly Review Press.

French, M. (1992) *The War against Women*, London: Hamish Hamilton.

Friedan, B. (1963) *The Feminine Mystique*, Harmondsworth: Pelican.

Gaitskell, D., Kimbie, J., Maconachie, M. and Unterhalter, E. (1984) 'Class, race and gender: domestic workers in South Africa', *Review of African Political Economy*, 27/28: 86–108.

Gamarnikow, E. *et al.* (eds) (1983) *Gender, Class and Work*, London: Heinemann.

Game, A. and Pringle, R. (1983) *Gender at Work*, Sydney: George Allen & Unwin.

Gardiner, J. (1975) 'Women's domestic labour', *New Left Review* 89: 47–71.

Garland, T. (1972) 'The better half? The male in the dual profession family', in C. Safilios-Rothschild (ed.) *Towards a Sociology of Women*, Lexington: Xerox College Publishing.

Gathorne-Hardy, J. (1972) *The Rise and Fall of the British Nanny*, London: Hodder & Stoughton.

Gershuny, J.I. (1978) *After Industrial Society? The Emerging Self-Service Economy*, Basingstoke: Macmillan.

—— (1983) *Social Innovation and the Division of Labour*, Oxford: Oxford University Press.

—— (1985) 'Economic development and change in the mode of provision of services', in N. Redclift and E. Mingione (eds) *Beyond Employment*, Oxford: Basil Blackwell.

Gershuny, J.I. and Miles, I.D. (1983) *The New Service Economy: the Transformation of Employment in Industrial Societies*, London: Frances Pinter.

Gill, L. (1988) 'Senora and Sirvientas: women and domestic services in La Paz, Bolivia', paper presented to 14th International Congress of the Latin American Studies Association, New Orleans.

Gillespie, A. and Green, A. (1987) 'The changing geography of "producer services" employment in Britain', *Regional Studies* 21, 5: 397–411.

Gillis, J.R. (1983) 'Servants, sexual relations and the risks of illegitimacy in London, 1801–1900', in J.L. Newton, M.P. Ryan and J.R. Walkowitz (eds) *Sex and Class in Women's History*, London: Routledge & Kegan Paul.

Glendinning, C. (1991) 'Dependency and interdependency: the incomes of informal carers and the impact of social security', *Journal of Social Policy* 19, 4: 469–97.

Glendinning, C. and Millar, J. (eds) (1989) *Women and Poverty in Britain*, Brighton: Wheatsheaf.

Glenn, E. (1980) 'The dialectics of wage work: Japanese-American women and domestic service, 1905–1940', *Feminist Studies* 6, 3: 432–71.

—— (1981) 'Occupational ghettoisation: Japanese-American women and domestic service, 1905–1970', *Ethnicity* 8, 4: 352–86.

—— (1986) *Issei, Nisei, War Bride: Three Generations of Japanese-American Women in Domestic Service*, Philadelphia: Temple University Press.

—— (1988) 'A belated industry revisited: domestic service amongst Japanese-American women', in A. Stratham *et al.* (eds) *The Worth of Women's Work: A Qualitative Synthesis*, Albany: State University of New York Press.

—— (1992) 'From servitude to service work: historical continuities in the racial division of paid reproductive labour', *Signs*, 18, 1: 1–43.

Glucksmann, M. (1990) *Women Assemble: Women Workers and the New Industries in Inter-War Britain*, London: Routledge.

Gogna, M. (1989) 'Domestic workers in Buenos Aires', in E. Chaney and M. Garcia Castro (eds) *Muchachas No More*, Philadelphia: Temple University Press.

Goldthorpe, J. (1982) 'On the service class: its formation and future', in A. Giddens and G. Mackenzie (eds) *Social Class and the Division of Labour*, Cambridge: Cambridge University Press.

—— (1983) 'Women and class analysis: in defence of the conventional view', *Sociology* 17, 4: 465–88.

—— (with C. Llewellyn and C. Payne) (1980) *Social Mobility and the Class Structure in Modern Britain*, Oxford: Clarendon Press.

Goode, W. (1960) 'A theory of role strain', *American Sociological Review* 25, 4: 483–96.

Gorz, A. (1989) *Critique of Economic Reason*, London: Verso.

Graham, H. (1983) 'Caring: a labour of love', in J. Finch and D. Groves (eds) *A Labour of Love: Women, Work and Caring*, London: Routledge & Kegan Paul.

—— (1991) 'The concept of caring in feminist research: the case of domestic service', *Sociology* 25, 1: 61–78.

Grant, L. *et al.* (1990) 'Gender, parenthood and the work hours of physicians', *Journal of Marriage and the Family* 52: 39–49.

Green, J. (1982) 'A survey of domestic service', *Lincolnshire History and Archaeology* 17: 65–9.

Gregson, N. and Lowe, M. (1989) 'Nannies, cooks, cleaners, au pairs . . . New issues for feminist geographers', *Area* 21, 4: 415–17.

—— and —— (1993) 'Re-negotiating the domestic division of labour? A study of dual-career households in north-east and south-east England', *Sociological Review* 41,3: 475–505.

—— and —— (forthcoming a) 'Waged domestic labour and the renegotiation of the domestic division of labour within dual-career households', *Sociology*.

—— and —— (forthcoming b) 'Waged domestic labour and the construction of new forms of femininity'.

—— and —— (forthcoming c) 'Whose representations, whose empowerment? Thoughts on the power relations of social research'.

Griffin, C. (1985) *Typical Girls? Young Women from School to the Jobmarket*, London: Routledge & Kegan Paul.

Gross, S. (1991) 'Domestic labour as a life course event: the effects of ethnicity in turn of the century America', *Social Science History* 15, 3: 397–416.

Hagen, E. and Jenson, J. (1988) 'Paradoxes and promises: work and politics in the post war years', in J. Jenson, E. Hagen and C. Reddy (eds) *Feminisation of the Labour Force: Paradoxes and Promises*, Cambridge: Polity.

Halford, S. and Savage, M. (forthcoming) 'Charging masculinities: management and careers'.

—— Savage, M. and Witz, A. (forthcoming) *Gender, Careers and Organisations*.

Hall, P. Breheny, M., McQuaid, R. and Hart, D. (1987) *Western Sunrise: the Genesis and Growth of Britain's Major Hi-Tech Corridor*, London: Allen & Unwin.

Hamnett, C. (1976) 'Social change and social segregation in Inner London, 1961–71', *Urban Studies* 13: 261–71.

—— (1986) 'The changing socio-economic structure of London and the south-east', *Regional Studies* 20, 5: 391–406.

—— (1989) 'Consumption and class in contemporary Britain', in C. Hamnett, L. McDowell and P. Sarre (eds) *The Changing Social Structure*, London: Sage.

—— (1991) 'Labour markets, housing markets and social restructuring in a global city: the case of London', in J. Allen and C. Hamnett (eds) *Housing and Labour Markets*, London: Unwin Hyman.

—— and Williams, P. (1980) 'Social change in London: a study of gentrification', *Urban Affairs Quarterly* 15: 469–87.

Harris, C.C. (1985) *Redundancy and Recession*, Oxford: Basil Blackwell.

Harrison, R. (1975) *Rose: My Life in Service*, New York: Viking.

Hartmann, H. (1981a) 'The unhappy marriage of marxism and feminism: towards a more progressive union', in L. Sargent (ed.) *Women and Revolution*, London: Pluto.

—— (1981b) 'The family as the locus of gender, class and political struggle: the example of housework', *Signs* 6, 3: 366–94.

Hecht, J. (1956) *The Domestic Servant Class in Eighteenth-Century England*, London: Routledge & Kegan Paul.

Hertz, R. (1986) *More Equal than Others: Women and Men in Dual Career Marriages*, Berkeley: University of California Press.

Higgins, A. (1991) 'Getting clean away', *Good Housekeeping* (March).

Higgs, E. (1983) 'Domestic servants and households in Victorian England', *Social History* 8, 2: 201–10.

—— (1986a) *Domestic Servants and Households in Rochdale, 1851–1871*, New York: Garland.

—— (1986b) 'Domestic service and household production', in A. John (ed.) *Unequal Opportunities: Women's Employment in England, 1800–1918*, Oxford: Basil Blackwell.

Hochschild, A. (1989) *The Second Shift: Working Parents and the Revolution at Home*, New York: Viking.

Holmstrom, L.L. (1972) *The Two Career Family*, Cambridge, Mass.: Schenkman Publishing.

Horn, P. (1975) *The Rise and Fall of the Victorian Servant*, New York: St Martin's Press.

Huggett, F.E. (1977) *Life Below Stairs: Domestic Servants in England from Victorian Times*, New York: Scribner.

Hunt, J. and Hunt, L. (1977) 'Dilemmas and contradictions of status: the case of the dual-career family', *Social Problems* 24: 407–16.

—— and —— (1982) 'The dualities of careers and families: new integrations or new polarisations', *Social Problems* 29: 499–510.

Jamieson, L. (1990) 'Rural and urban women in domestic service', in E. Gordon and E. Breitenbach (eds) *The World is Ill Divided*, Edinburgh, Edinburgh University Press.

Jelin, E. (1977) 'Migration and labour force participation of Latin American women: the domestic servants in the cities', in Wellesley Editorial Committee (eds) *Women and National Development*, Chicago: Chicago University Press.

Johnson, C. and Johnson, F.A. (1980) 'Parenthood, marriage and careers: situational constraints and role strain', in F. Pepitone-Rockwell (ed.) *Dual Career Couples*, Beverley Hills: Sage.

Katzman, D.M. (1978) *Seven Days a Week: Women and Domestic Service in Industrialising America*, New York: Oxford University Press.

Kessler Harris, A. (1982) *Out to Work: a History of Wage-earning Women in the United States*, New York: Oxford University Press.

Land, H. (1978) 'Who cares for the family?' *Journal of Social Policy* 7, 3: 257–84.

Lash, S. and Urry, J. (1987) *The End of Organised Capitalism*, Cambridge: Polity.

Laslett, B. and Brenner, J. (1989) 'Gender and social reproduction', *American Review of Sociology*, 15: 381–404.

Laslett, P. (1965) *The World We Have Lost*, London: Methuen.

Laughlin, G. (1901) 'Domestic service', in *Report and Testimony of the Industrial Commission Volume XIV*, Washington DC.

Lawe, C. and Lawe, B. (1980) 'The balancing act: coping strategies for emerging family lifestyles', in F. Pepitone-Rockwell (ed.) *Dual Career Couples*, Beverley Hills: Sage.

Leadbetter, C. (1989) 'Boomtown blues', *Marxism Today*, October 20–23.

Leashore, B.R. (1984) 'Black female workers: live-in domestics in Detroit, Michigan (1860–1880)', *Phylon* 45, 2: 111–20.

Lewis, J. (1980) *The Politics of Motherhood: Child and Material Welfare in England, 1900–1939*, London: Croom Helm.

—— (1983) 'Women, work and regional development', *Northern Economic Review* 7: 10–24.

—— (ed.) (1983) *Women's Welfare, Women's Rights*, London: Croom Helm.

—— (1984) *Women in England, 1870–1950*, Brighton: Wheatsheaf.

—— (ed.) (1986) *Labour and Love: Women's Experience of Home and Family, 1850–1940*, Oxford: Basil Blackwell.

—— and Meredith, B. (1988) *Daughters Who Care*, London: Routledge.

Lewis, S. and Cooper, C.L. (1989) *Career Couples: Contemporary Lifestyles and How to Manage Them*, London: Unwin.

Leyshon, A. and Thrift, N. (1989) 'South goes North? The rise of the British provincial financial centre', in J.R. Lewis and A.R. Townsend (eds) *The North/South Divide*, London: Paul Chapman.

Lintelman, J. (1991) '"Our serving sisters": Swedish-American domestic servants and their ethnic community', *Social Science History* 15, 3: 381–96.

Litvinoff, S. and Velmans, M. (1987) *Working Mother: A Practical Handbook*, London: Transworld Publishers.

Lockwood, D. (1986) 'Class, status and gender', in R. Crompton and M. Mann (eds) *Gender and Stratification*, Cambridge: Polity.

McBride, T.M. (1974) 'Social mobility for the lower classes: domestic servants in France', *Journal of Social History* 8: 63–78.

—— (1976) *The Domestic Revolution: the Modernisation of Household Service in England and France, 1820–1920*, London: Croom Helm.

—— (1978) '"As the twig is bent": the Victorian nanny', in A.S. Wohl (ed.) *The Victorian Family: Structure and Stresses*, London: Croom Helm.

315

McCrindle, J. and Rowbotham, S. (eds) (1977) *Dutiful Daughters*, Harmondsworth, Pelican.

McDowell, L. (1989) 'Women in Thatcher's Britain', in J. Mohan (ed.) *The Political Geography of Contemporary Britain*, London: Macmillan.

—— (1991) 'Life without father and Ford: the new gender order of post-Fordism', *Transactions of the Institute of British Geographers* 16, 4: 400–19.

—— and Massey, D. (1984) 'A Woman's Place?' in J. Allen and D.B. Massey (eds) *Geography Matters*, Milton Keynes: Open University Press.

Mack, J. and Lansley, S. (1985) *Poor Britain*, London: George Allen & Unwin.

Mackenzie, S. (1989) *Visible Histories: Women and Environments in a Post War British City*, Montreal: McGill–Queen's University Press.

—— and Rose, D. (1983) 'Industrial changes, the domestic economy and home life', in J. Anderson, S. Duncan and R. Hudson (eds) *Redundant Spaces: Industrial Decline in Cities and Regions*, London: Academic Press.

McNally, F. (1979) *Women for Hire: A Study of the Female Office Worker*, Basingstoke: Macmillan.

Macpherson, C.B. (1973) 'Servants and labourers in seventeenth-century England', in *Democratic Theory: Essays in Retrieval*, Oxford: Clarendon Press.

McRobbie, A. (1978) 'Working-class girls and the culture of femininity', in Centre for Contemporary Cultural Studies, *Women Take Issue: Aspects of Women's Subordination*, London: Hutchinson.

Malos, E. (1980) *The Politics of Housework*, London: Allison & Busby.

Mandel, (1975) *Late Capitalism*, London: Verso.

Marshall, D. (1949) *The English Domestic Servant in History*, London: Historical Association.

Marshall, G., Newby, H., Rose, D. and Vagler, C. (1988) *Social Class in Modern Britain*, London: Unwin Hyman.

Marshall, N. (1985) 'Research policy and review 4: services in a post industrial economy', *Environment and Planning* A 17: 1155–67.

——, Damesick, P.J. and Wood, P. (1987) 'Understanding the location and role of producer services in the UK', *Environment and Planning* A 19: 575–95.

Martin, J. and Roberts, C. (1984) *Women and Employment: A Lifetime Perspective*, London: HMSO.

Martin, R. (1988) 'Industrial capitalism in transition: the contemporary re-organisation of the British space economy', in D.B. Massey and J. Allen (eds) *Uneven Redevelopment*, London: Hodder & Stoughton.

Massey, D.B. (1984) *Spatial Divisions of Labour*, London: Macmillan.

—— (1988) 'Uneven redevelopment: social change and spatial divisions of labour', in D.B. Massey and J. Allen (eds) *Uneven Redevelopment*, London: Hodder & Stoughton.

Mattingly, D.J. (1993) 'Home, work and domestic service on the US–Mexico border', paper presented to the AAG Conference, Atlanta.

Maza, S. (1983) *Servants and Masters in Eighteenth-Century France: the Uses of Loyalty*, Princeton, Princeton University Press.

Melhuish, E. and Moss, P. (eds) (1990) *Day Care for Young Children: International Perspectives*, London: Routledge, Chapman & Hall.

Moen, P. and Dempster McClain, D.I. (1987) 'Employed parents, role strain, work time and preferences for working less', *Journal of Marriage and the Family* 49: 579–90.

Morris, L.D. (1985a) 'Local social networks and domestic organisations: a study of redundant steelworkers and their wives', *Sociological Review* 33 (2) 327–42.

—— (1985b) 'Re-negotiation of the domestic division of labour in the context of male redundancy', in B. Roberts *et al.* (eds) *New Approaches to Economic Life*, Manchester: Manchester University Press.

—— (1987) 'Constraints on gender: the family wage, social security and the labour market: reflections on research in Hartlepool', *Work, Employment and Society* 1, 1: 85–106.

—— (1988) 'Employment, the household and social networks', in D. Gallie (ed.) *Employment in Britain*, Oxford: Basil Blackwell.

—— (1989) 'Household strategies: the individual, the collectivity and the labour market: the case of married couples', *Work, Employment and Society* 3: 447–64.

—— (1990) *The Workings of the Household: a US–UK comparison*, Cambridge: Polity.

—— (1991) 'Locality studies and the household', *Environment and Planning A* 23: 165–77.

—— and Irwin, S. (1992) 'Employment histories and the concept of the underclass', *Sociology* 26, 3: 401–20.

Mortimer, J., Hall, R. and Hill, R. (1976) 'Husbands' occupational attitudes as constraints on wives' employment', Paper presented at annual meeting of the American Sociological Association.

Moss, P. (1986) 'Some principles for a childcare service for working parents', in Equal Opportunities Commission, *Childcare and Equal Opportunities: Some Policy Perspectives*, London: HMSO.

Mottershead, P. (1988) *Recent Developments in Childcare: A Review*, London: HMSO, Equal Opportunities Commission.

Munt, I. (1987) 'Economic restructuring, culture and gentrification: a case study in Battersea, London', *Environment and Planning A* 19: 1175–97.

Myrdal, A. and Klein, V. (1956) *Women's Two Roles*, London: Routledge & Kegan Paul.

Neff, W.F. (1966) 'The governess', in *Victorian Working Women: an Historical and Literary Study of Women in British Industries and Professions, 1832–1850*, New York: AMS Press.

Nett, E.M. (1966) 'The servant class in a developing country: Ecuador', *Journal of Inter-American Studies* 8, 3: 437–52.

Newton, J.L., Ryan, M.P. and Walkowitz, J.R. (eds) (1983) *Sex and Class in Women's History*, London: Routledge & Kegan Paul.

Oakley, A. (1974) *The Sociology of Housework*, Oxford: Martin Robertson.

—— (1990) *Housewife*, Harmondsworth: Penguin.

Owens, C. and Moss, P. (1989) 'Patterns of pre-school provision in English local authorities', *Journal of Education Policy* 4, 4: 309–28.

Pahl, R. (1984) *Divisions of Labour*, Oxford: Basil Blackwell.

—— (ed.) (1988) *On Work*, Oxford: Basil Blackwell.

Palmer, P. (1987) 'Housewife and household worker: employer–employee relations in the home, 1928–41', in C. Groneman and M. Norton (eds) *To Toil the Livelong Day*, Ithaca, New York: Cornell University Press.

—— (1990) *Domesticity and Dirt: Housewives and Domestic Servants in the US, 1920–45*, Philadelphia: Temple University Press.

Pendleton, B.F. *et al.* (1982) 'An approach to quantifying the needs of dual-career families', *Human Relations* 35, 1: 69–82.

Pepitone-Rockwell, F. (ed.) (1980) *Dual Career Couples*, Beverly Hills: Sage.

Peterson, J.J. (1972) 'The Victorian governess: status incongruity in family and society', in M. Vicinus (ed.) *Suffer and Be Still*, Bloomington: Indiana University Press.

Pettengill, L. (1903) *Toilers of the Home: A Report of a College Woman's Experience as a Domestic Servant*, New York: Doubleday.

Phillips, A. (1987) *Divided Loyalties*, London: Virago.

Phillips, C. (1991) 'You just can't get the staff these days', *Evening Standard* (18 March).

Phizacklea, A. (1982) 'Migrant women and wage labour. The case of West Indian women in Britain', in J. West (ed.) *Work, Women and the Labour Market*, London: Routledge & Kegan Paul.

—— (ed.) (1983) *One Way Ticket*, London: Routledge & Kegan Paul.

—— (1987) 'Minority women and economic restructuring: the case of Britain and the Federal Republic of Germany', *Work, Employment and Society* 1, 1: 309–25.

Pinch, S. (1987) 'Labour market theory, quantification and policy', *Environment and Planning A*, 19: 1477–94.

Pleck, J. (1985) *Working Wives/Working Husbands*, Beverley Hills: Sage.

Podmore, D. and Spencer, A. (1986) 'Gender in the labour process: the case of men and women lawyers', in D. Knights and H. Willmott (eds) *Gender and the Labour Process*, Aldershot: Gower.

Pollert, A. (1981) *Girls, Wives, Factory Lives*, London: Macmillan.

Poloma, M. (1972) 'Role conflict and the married professional woman', in C. Safilios-Rothschild (ed.) *Towards a Sociology of Women*, Lexington: Xerox College Publishing.

—— and Garland, T. (1971) 'The married professional woman: a study of the tolerance of domestication', *Journal of Marriage and the Family* 33: 531–40.

Pond, C. (1989) 'The changing distribution of income, wealth and poverty', in C. Hamnett, L. McDowell and P. Sarre (eds) *The Changing Social Structure*, London: Sage.

Preston Whyte, E. (1976) 'Race attitudes and behaviour: the case of domestic employment in white South African homes', *African Studies* 35, 2: 71–89.

Pringle, R. (1990) *Secretaries Talk*, London: Verso.

Prochaska, F.K. (1981) 'Female philanthropy and domestic service in Victorian England', *Bulletin of the Institute of Historical Research* 54, 129: 78–85.

Radcliffe, S. (1990) 'Ethnicity, patriarchy and incorporation into the nation: female migrants as domestic servants in Peru', *Environment and Planning D: Society and Space* 8, 4: 379–93.

Rajan, A. (1987) *Services: the Second Industrial Revolution?* London: Butterworth.

Ramazanoglu, C. (1989) *Feminism and the Contradictions of Oppression*, London: Routledge.

Rapoport, R. and Rapoport, R.N. (1969) 'The dual earner family: a variant pattern and social change', *Human Relations* 22, 4: 3–30.

—— and —— (1971) *Dual Career Families*, Harmondsworth: Penguin.

—— and —— (1972) 'The dual-career family: a variant pattern and social change', in C. Safilios-Rothschild (ed.) *Towards a Sociology of Women*, Lexington: Xerox College Publishing.

—— and —— (1978) (eds) *Working Couples*, New York: Harper & Row.

—— and —— (1980) 'Three generations of dual-career family research', in F. Pepitone-Rockwell (ed.) *Dual Career Couples*, Beverly Hills: Sage.

Rapoport, R.N. *et al.* (eds) (1982) *Families in Britain*, London: Routledge & Kegan Paul.

Rennie, J. (1982) *Every Other Sunday*, New York: St Martin's Press.

Riley, D. (1979) *War in the Nursery: Theories of the Child and Mother*, London: Virago.

Roberts, E. (1984) *A Woman's Place: An Oral History of Working Class Women, 1890–1940*, Oxford: Basil Blackwell.

—— (1988) *Women's Work, 1840–1940*, Basingstoke: Macmillan.

Robinson, F. (ed.) (1988) *Post Industrial Tyneside*, Newcastle upon Tyne: Newcastle upon Tyne City Libraries and Arts.

—— and Gregson, N. (1992) 'The, "underclass": a class apart', *Critical Social Policy* 34, Summer: 38–51.

Robinson, M. (1924) *Domestic Workers and their Employment Relations*, Washington: Women's Bureau Bulletin, no. 39.

Rollins, J. (1985) *Between Women: Domestics and their Employers*, Philadelphia: Temple University Press.

Romero, M. (1988) 'Day work in the suburbs: the work experience of Chicana private housekeepers' in A. Stratham, E. Miller and H. Mauksch (eds) *The Worth of Women's Work: A Qualitative Synthesis*, Albany: State University of New York Press.

Root, A. (1984) 'The return of the nanny', *New Socialist* (December).

Rose, N.B., Jerdee, T.H. and Prestwich, T.L. (1975) 'Dual career marital adjustment: potential effects of discriminatory managerial attitudes', *Journal of Marriage and the Family*, 37: 565–72.

Rostow, W.W. (1960) *The Stages of Economic Growth: a Non-Communist Manifesto*, Cambridge: Cambridge University Press.

Rowbotham, S. (1977) *Hidden from History: 300 Years of Women's Oppression and the Fight against it*, London: Pluto.

Rubbo, A. and Taussig, M. (1978) 'Up off their knees: servanthood in south west Colombia', in *Female Servants and Economic Development*, Occasional Papers in Women's Studies, no. 1, Ann Arbor: Michigan University.

Rubery, J. (ed.) (1988) *Women and Recession*, London: Routledge & Kegan Paul.

—— and Humphries, J. (1988) 'Recession and exploitation: British women in a changing workplace', in J. Jenson *et al.* (eds) *Feminisation of the Labour Force: Paradoxes and Promises*, Cambridge: Polity.

—— and Tarling, R. (1988) 'Women's employment in declining Britain', in J. Rubery (ed.) *Women and Recession*, London: Routledge & Kegan Paul.

—— and —— (eds) (1988) *Women and Recession*, London: Routledge.

Salmon, L.M. (1897) *Domestic Service*, New York: Macmillan.

Savage, M. (1985) 'Capitalist and patriarchal relations at work: Preston cotton weaving, 1890–1940', in Lancaster Regionalism Group, *Localities, Class and Gender*, London: Pion.

—— (1988) 'The missing link? The relationship between spatial mobility and social mobility', *British Journal of Sociology* 39, 4: 554–77.

—— and Fielding, A.J. (1989) 'Class formation and regional development: the "service class" in south east England', *Geoforum* 20, 2: 203–18.

—— and Witz, A. (eds) (1992) *Gender and Bureaucracy*, Oxford: Blackwell/ *Sociological Review*.

——, Barlow, J., Dickens, P. and Fielding, T. (1992) *Property, Bureaucracy and Culture: Middle-Class Formation in Contemporary Britain*, London: Routledge.

——, Dickens, P. and Fielding, A.J. (1988) 'Some social and political implications of the contemporary fragmentation of the service class, *International Journal of Urban and Regional Research* 12, 3: 455–76.

Sayer, A. (1984) *Method in Social Science: A Realist Approach*, London: Hutchinson.

Seccombe, W. (1974) 'The housewife and her labour under capitalism', *New Left Review* 83: 3–24.

319

Segal, L. (1987) *Is the Future Female? Troubled Thoughts on Contemporary Feminism*, London: Virago.

Sharpe, S. (1984) *Double Identity: the Lives of Working Mothers*, Harmondsworth: Penguin.

Silverstone, R. (1980) 'Accountancy', in R. Silverstone and A. Ward (eds) *Careers of Professional Women*, London: Croom Helm.

Smith, M.L. (1973) 'Domestic service as a channel of upward mobility for the lower class woman: the Lima case', in A. Pescatello (ed.) *Female and Male in Latin America*, Pittsburgh: Pittsburgh University Press.

—— (1989) 'Where is Maria now? Former domestic workers in Peru', in E. Chaney and M. Garcia Castro (eds) *Muchachas No More*, Philadelphia: Temple University Press.

Spock, B. (1966) *Baby and Childcare*, New English Library Ltd, Giant Cardinal Edition.

Steiner, J.M. (1989) *How to Survive as a Working Mother*, London: Kogan Page.

Stone, K. (1983) 'Motherhood and waged work: West Indian, Asian and White Mothers compared', in A. Phizacklea (ed.) *One Way Ticket: Migration and Female Labour*, London: Routledge & Kegan Paul.

Summerfield, P. (1984) *Women Workers in the Second World War*, London: Croom Helm.

Sutherland, D.E. (1981) *Americans and their Servants: Domestic Service in the United States from 1800 to 1920*, Baton Rouge: Louisiana State University Press.

Taylor, J.G. (1979) *From Modernisation to Modes of Production: A Critique of Sociology of Development*, London: Macmillan.

Taylor, P. (1979) 'Daughters and mothers – maids and mistresses: domestic service between the wars', in J. Clarke, C. Crichter and R. Johnson (eds) *Working Class Culture: Studies in History and Theory*, London: Hutchinson.

Thrift, N. (1987) 'The geography of late twentieth-century class formation', in N. Thrift and P. Williams (eds) *Class and Space*, London: Routledge & Kegan Paul.

—— (1989) 'Images of social change', in C. Hamnett *et al.* (eds) *The Changing Social Structure*, London: Sage.

Tilly, L.A. Luton, S. and Sankar, A. (1978) *Female Servants and Economic Development*, Ann Arbor: Michigan Occasional Paper in Women's Studies, no. 1.

Tranberg Hansen, K. (1986a) 'Domestic service in Zambia', *Journal of Southern African Studies* 13, 1: 57–81.

—— (1986b) 'Sex and gender among domestic servants in Zambia', *Anthropology Today* 2, 3: 18–23.

Ungerson, C. (1983) 'Why do women care?', in J. Finch and D. Groves (eds) *A Labour of Love: Women, Work and Caring*, London: Routledge & Kegan Paul.

—— (1987) *Policy is Personal: Sex, Gender and Informal Care*, London: Tavistock.

Urry, J. (1987) 'Some social and spatial aspects of services', *Environment and Planning D: Society and Space* 5, 1: 5–26.

—— (1990) *The Tourist Gaze*, London: Sage.

Van Onselen, C. (1982) 'The witches of suburbia: domestic service on the Witwatersrand, 1890–1914', in *Studies in the Social and Economic History of the Witwatersrand, 1866–1914*, New York: Longman.

Vax Home Cleaning Report (1991) *Keeping Britain's Homes Clean in the 1990s*.

Vicinus, M. (ed.) (1972) *Suffer and Be Still: Women in the Victorian Age*, Bloomington: Indiana University Press.

Wacjman, J. (1983) *Women in Control: Dilemmas of a Workers' Cooperative*, New York: St Martin's Press.

Walby, S. (1986) *Patriarchy at Work*, Cambridge: Polity.

—— (1990) *Theorising Patriarchy*, Cambridge: Polity.

Walker, A. (ed.) (1982) *Community Care: the Family, the State and Social Policy*, Oxford: Martin Robertson/Basil Blackwell.

Wallerstein, I. (1979) *The Capitalist World Economy*, Cambridge: Cambridge University Press.

Wallman, S. (1983) *Eight London Households*, London: Tavistock.

Wallston, B. *et al.* (1978) 'I will follow him – myth, reality or forced choice – job seeking experiences of dual-career couples', in J. Bryson and R. Bryson (eds) *Dual Career Couples*, Special Issue of *Psychology of Women Quarterly*, New York: Human Sciences Press.

Watson, S. (1991) 'The restructuring of work and home: productive and reproductive relations', in J. Allen and C. Hamnett (eds) *Housing and Labour Markets*, London: Unwin Hyman.

Weiner, L.Y. (1985) *'From Working Girl to Working Mother: the Female Working Force in the United States, 1820–1980'*, Berkeley: University of California Press.

Westwood, S. (1983) *All Day Every Day: Factory and Family in the Making of Women's Lives*, London: Pluto.

Wheelock, J. (1990) *Husbands at Home: the Domestic Economy in a Post Industrial Society*, London: Routledge.

Whisson, M.G. and Weil, W. (1971) *Domestic Servants: a Microcosm of the 'Race Problem'*, Johannesburg: South African Institute of Race Relations.

Willis, P. (1979) *Learning to Labour*, Farnborough: Saxon House.

Witz, A. (1990) 'Patriarchy and professions: the gendered politics of occupational closure', *Sociology* 24, 4: 675–90.

—— (1992) *Professions and Patriarchy*, London: Routledge.

Wolch, J. and Dear, M. (eds) (1989) *The Power of Geography*, London: Unwin Hyman.

Wood, P. (1986) 'The anatomy of job loss and job creation: some speculations on the role of the "producer service" sector', *Regional Studies* 20, 1: 37–46.

Working Mothers Association (1991) *The Working Parents Handbook*.

Wright, E.O. (1985) *Classes*, London: Verso.

Yeandle, S. (1984) *Women's Working Lives: Patterns and Strategies*, London: Tavistock.

Young, G.E. (1987) 'The myth of being "like a daughter"', *Latin American Perspectives* 54, 14: 365–80.

Young, M. and Willmott, P. (1973) *The Symmetrical Family*, New York: Penguin.

NAME INDEX

SUBJECT INDEX